KV-475-293

Introduction to
Chromatography

90 0680670 5

29 NOV 198

Charles Seale-Hayne Library
University of Plymouth
(01752) 588 588
LibraryandITenquiries@plymouth.ac.uk

WITHDRAWN
FROM
UNIVERSITY OF PLYMOUTH
LIBRARY SERVICES

Introduction to Chromatography

James M. Bobbitt
PROFESSOR OF CHEMISTRY
THE UNIVERSITY OF CONNECTICUT
STORRS, CONNECTICUT

Arthur E. Schwarting
PROFESSOR OF PHARMACOGNOSY
SCHOOL OF PHARMACY
THE UNIVERSITY OF CONNECTICUT
STORRS, CONNECTICUT

Roy J. Gritter
GROUP LEADER IN ANALYTICAL CHEMISTRY
RESEARCH DIVISION
INTERNATIONAL BUSINESS MACHINES, INC.
SAN JOSE, CALIFORNIA

D. Van Nostrand Company
New York • Cincinnati • Toronto • London • Melbourne

PLYMOUTH POLYTECHNIC
LEARNING RESOURCES CENTRE

| ACCN. No. | 76746 |
| CLASS No. | 544.92 BOB |

-0. SEP. 1977

D. Van Nostrand Company Regional Offices:
New York Cincinnati Millbrae

D. Van Nostrand Company International Offices:
London Toronto Melbourne

Copyright © 1968 by LITTON EDUCATIONAL PUBLISHING, INC.

Library of Congress Catalog Card Number 68-22841
ISBN: 0-442-17595-7

*All rights reserved. No part of this work covered by
the copyright hereon may be reproduced or used in
any form or by any means—graphic, electronic, or
mechanical, including photocopying, recording,
taping, or infomation storage and retrieval systems—
without written permission of the publisher.
Manufactured in the United States of America.*

Published by D. Van Nostrand Company
450 West 33rd Street, New York, N. Y. 10001

8 7 6 5 4 3

Preface

The laboratory practice of chemistry, either in itself or subsidiary to other fields, is becoming ever more dependent upon chromatography as an ultimate method for the separation of mixtures. Chemists investigate reaction mixtures chromatographically to learn what occurs during the reaction; they use chromatographic techniques to isolate and purify the products; and finally, they quantify the results chromatographically to gain insight into the course of reaction. Biologists use chromatographic methods to separate and identify the compounds of tissues. In drug-plant research and pharmacognosy, the processes of drug extraction, isolation, and purification are monitored from beginning to end by chromatography. In industry, the introduction of chromatographic methods for routine quality control has allowed the saving of vast amounts of time and money.

It thus appears that a book which would give a practical introduction to the more common techniques in this area and which would serve only as a point of departure would be highly desirable. This book should provide the undergraduate or beginning graduate student with basic knowledge of the field without overwhelming him with a multitude of permutations and variations. Furthermore, the book should be short and reasonably priced. We hope this text will be appropriate.

There are a great many chromatographic techniques which might well be included in such an introductory text. Due to the size and type of this book as well as our personal experience (and prejudice), it has been decided to limit the discussion to three techniques; thin-layer, column, and gas chromatography. Thus, one has two excellent qualitative methods, thin-layer and gas chromatography; one excellent quantitative technique, gas chromatography; and the classical preparative method, column chromatography. The major casualty of this selection is paper chromatography, a powerful and well-known technique. It is more than likely, however, that the considerable advantages of thin-layer over paper chromatography will, in time, lead to a substitution for the majority of the paper chromatography now used.

A number of outstanding books have been written on chromatography in general and on the several specific varieties. Most of these are quite complete, carefully referenced, and well written and organized. Most of them are also expensive and are normally not written for the beginner. In this text, the complete picture will not be given for each technique discussed, but various books which do will be cited in the bibliography. Except in the case of direct quotations, no references will be given. At the end of this book is an appendix containing information on available commercial chemicals and equipment.

In summary, enough information will be given in this text to introduce the subject and to permit the beginning student to solve some problems. We have drawn on our own personal experience in deciding what to present and apologize in advance to those who may not agree. Furthermore, we have tried to produce some simple answers to questions which really have no simple answers. We hope the reader will realize this and respect our generalizations as such and no more.

We would like to acknowledge the assistance of Mrs. Jane Ann Hickman Bobbitt who read and listened to the manuscript many times, Dr. H. K. Mangold of the Hormel Institute and Professor M. E. Morgan of the University of Connecticut who read portions of the manuscript, of Mrs. Adah Ruth Ballard who typed the manuscript with skill and forbearance, of Mr. Roland L. Laramie of Dineen Studios in Willimantic, Connecticut, who made many of the photographs, and finally, of our students who have aided, abetted, and tolerated our obsessions with chromatography.

JAMES M. BOBBITT
Storrs, Connecticut

ARTHUR E. SCHWARTING
Storrs, Connecticut

ROY J. GRITTER
San Jose, California

April, 1968

Contents

1

Introduction

Chromatography is a separation technique of great resolving power and considerable complexity. Although the word implies color, there is no direct connection except that the first compounds to be separated by the technique were plant pigments.

Chromatographic separations are carried out by ingenious mechanical manipulations involving a few of the general physical properties of molecules. The major properties involved are: (a) the tendency of a molecule to dissolve in a liquid (**solubility**); (b) the tendency for a molecule to attach itself to a finely divided solid (**adsorption**); and (c) the tendency for a molecule to enter the vapor state or evaporate (**volatility**). In chromatography, mixtures of substances to be separated are placed in a dynamic or moving experimental situation where they can exhibit *two* of these properties. This may involve using the same property twice, such as solubility in two different liquids, or it may involve two different properties entirely.

The nature of these ideas can best be shown by describing a few static examples of the interactions between two properties. Thus, if one places a substance in a separatory funnel with two liquids which have a limited mutual solubility (such as ether and water) the substance will tend to distribute itself or to **partition** itself between the two liquids, depending upon its relative solubility in them. This is an interaction between the solubility in two liquids, or the first property used twice. If one places a substance in a flask with a liquid and a finely divided solid (such as charcoal), the substance will distribute itself between the liquid and the surface of the solid. Thus, the solubility and the adsorptive properties are used together. Finally, if a volatile substance is dissolved in a nonvolatile liquid which is present in a thin film, it will distribute itself between the liquid film and the gas in contact with the film. This represents an interplay

1

between the solubility and the volatility properties. There are, of course, several other possibilities. It is necessary that each of the components of the above systems be in intimate contact and in as complete an equilibrium as possible. The systems can be said to consist of two **phases** with a substance distributed between them.

It is highly unlikely that two different substances will exhibit, quantitatively, exactly the same pair of physical properties. It is these differences, which may be very small indeed, that serve as the basis for chromatographic separations.

The application of these principles to the practical separation of mixtures requires considerable mechanical ingenuity. The basic idea is to convert, experimentally, one of the static situations described previously into a dynamic or moving situation. Thus, one of the phases (the **moving phase**) is caused to flow past the other phase (the **stationary phase**) while still remaining, at least to some extent, in equilibrium with it. The moving phase can be a liquid or a gas and the stationary phase can be a liquid film or a finely divided solid. The mixture is then introduced into the system. The properties of the mixture components will determine whether they move or not, and if they do, how fast they move relative to one another and the moving phase. Since separation would be impossible if nothing moved at all, the phases are generally chosen such that all of the components of the mixture move, but at varying rates. This difference in migration rates is the basis for chromatographic separations.

If the moving phase is a liquid and the stationary phase is a liquid film on some suitable support, the resulting chromatography is called **partition chromatography.** If the moving phase is a liquid and the stationary phase is a solid functioning as an adsorbing surface (rather than as a support for a liquid film) the result is **adsorption chromatography.** In the situations where the moving phase is a gas, the stationary phase is almost always a liquid film. Thus, a **gas-liquid-partition chromatography** results. In the rare cases where the stationary phase is an adsorbing surface, **gas-solid-adsorption chromatography** is concerned. The last two types are generally referred to simply as **gas chromatography.**

It is not always possible to prescribe exactly the type of chromatography that will prevail or to describe which type actually yielded a separation. In many cases, at least two phenomena (generally adsorption and partition) are taking place.

There are three additional types of chromatography which will not be considered in this book, but which should be mentioned. These are **ion-exchange chromatography, electron-exchange chromatography** and **gel-filtration chromatography.** In all cases, the experimental techniques and the observed results are the same as in partition or adsorption chromatography. However, the underlying principles are different.

In ion-exchange chromatography, the stationary phase is an acid or base, normally consisting of an acidic functional group (such as a carboxylic acid or a sulfonic acid) or a basic functional group (an amine or a quaternary amine hydroxide) present on a polymer chain. The polymeric nature renders the substance insoluble in normal solvents, including water, but still allows the functional groups to be operative. The moving phase is an ionic material, such as an acid, a base, or a salt, dissolved in water and so chosen that it can compete with the stationary phase for the substances to be separated.

The stationary phase in electron-exchange chromatography is an oxidizing or reducing agent, again polymeric in nature so that it is not soluble in normal solvents. The moving phase is an oxidizing or reducing agent chosen so that it can compete with the stationary phase.

Gel filtration involves chromatography on a substance of known pore size or porosity such as a cross-linked polydextran (Sephadex). The solutes may be ionic or nonionic. The adsorbent has the ability to exclude or provide accessible sites for molecules of specific sizes. Thus, the separation is based upon molecular size alone.

1.1. DEFINITION OF TERMS

The above introductory remarks have necessarily involved the definition of some important terms, but several more terms need to be defined in a general discussion of chromatography.

The solid material which serves as the stationary phase in adsorption chromatography is generally called the **adsorbent,** whereas the support for a liquid film in partition chromatography is generally called the **support.** When the moving phase is caused to flow over the adsorbent or support, the process is known as **development.** After the substances on the chromatogram have been separated by development, they are **detected** or **visualized.** If the substances on the chromatogram are actually washed off of the adsorbent, **elution** has taken place. The substances being separated are normally termed the **solutes** or, collectively, the **sample.**

In order to conserve space, two abbreviations will be used frequently. These are TLC for thin-layer chromatography, and GLC for gas liquid chromatography. Several other terms specific to these techniques will be defined later.

1.2. TECHNIQUES

Thin-Layer Chromatography

In TLC the adsorbent is deposited in a thin layer (0.1–2 mm thick) on a flat supporting surface. The supporting surface is normally a piece of glass and the adsorbent is generally held in place with a binding agent such

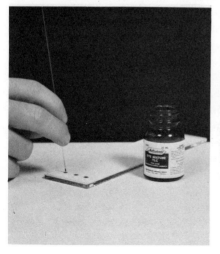

FIG. 1.1

FIG. 1.2

FIG. 1.3

FIG. 1.4

FIGS. 1.1 to 1.4. A mixture of three dyes (butter yellow, Sudan red III, and indophenol, Mallinckrodt Chem. Co.) in benzene is spotted on a silica gel G layer in three concentrations (Fig. 1.1); placed in a saturated chamber containing benzene as a developer (Fig. 1.2); developed (Fig. 1.3); and removed and dried (Fig. 1.4).

as starch or plaster of Paris. The mixture to be separated is dissolved in any suitable solvent and applied as a spot (3–5 mm in diameter) to the thin layer a short distance (2 cm) from one end. Such an application can be made with a capillary (Figure 1.1) or a syringe. The solvent is allowed to evaporate or is removed in a stream of warm air. The layer is then placed vertically in a developing chamber containing solvent in the bottom (about 1 cm deep) in such a manner that the solvent is in contact with the adsorbent layer on the end nearest the sample spot (Figure 1.2). The chamber is then closed and the solvent is allowed to ascend the layer by capillary action (Figure 1.3). The development is allowed to proceed until the solvent has climbed 10–15 cm above the point where the mixture was placed. If properly selected, the solvent will resolve the original spot of mixture into a series of spots, each, hopefully, corresponding to a single component (Figure 1.4). If the spots are not colored compounds as shown in the figures, they are visualized by spraying them with a suitable color-producing agent.

The chromatography is usually carried out in a chamber which has been as nearly saturated as possible with the solvent system that will be used in the development. This is the purpose served by the filter paper which partially lines the walls of the chambers shown in Figures 1.2 and 1.4.

Several terms in addition to those already given are frequently used in describing TLC (most of them actually coming from paper chromatography). The spot containing the mixture at the beginning of the chromatogram is called the **origin** and the technique of placing it there is known as **spotting**. The **solvent front** is the top of the layer of solvent as it flows up through the chromatogram, and, after the development is finished, represents the maximum height achieved by the solvent.

The behavior of a specific compound in a specific chromatographic system is frequently described by the R_f value. This number is obtained by dividing the distance moved by the solvent front into the distance moved by the compound (measured to the center of the spot). Both values are measured from the origin, and the R_f varies from zero to unity. This is graphically shown in Figure 1.5.

Column Chromatography

In column chromatography the stationary phase is placed in a cylindrical tube, usually made of glass, closed on the bottom with a valve or stopcock. This phase may be a finely divided adsorbent for adsorption chromatography or an inert material supporting a liquid film for partition chromatography. The column can be prepared with dry adsorbent or as a slurry (Figure 1.6). If the column is packed dry, it is wetted, after packing, with the liquid moving phase so that the prepared column in either case is in contact with a liquid.

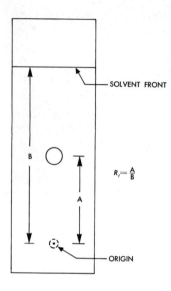

FIG. 1.5. An idealized thin-layer chromatogram showing how an R_f value is measured and calculated.

$$R_f = \frac{A}{B}$$

The liquid level is lowered to the top of the adsorbent, and the sample to be separated, dissolved in a minimum of solvent (generally the liquid already on the column), is placed on the column and allowed to flow in (Figure 1.7). A solvent or an appropriate mixture of solvents is then allowed to flow through the chromatogram to develop it. The liquid that flows out of the column is called the **effluent.** The solvent mixture, in contrast to TLC, is generally changed during the development, usually being made more polar. If the correct solvents and adsorbents have been chosen, the substances to be separated proceed down through the column (as **bands**) at differing rates (Figure 1.8) and eventually flow out in the effluent at different times (Figure 1.9).

The effluent can be continually analyzed as it emerges from the column or can be divided into fractions and analyzed (frequently by TLC) (Figure 4.5B). The portions of the effluent which contain the same substances are combined and the substances are isolated by solvent evaporation. This process is an analysis (of a mixture) by elution and is called **elution analysis.** If the solvent is changed during the development, the process is called **gradient elution analysis.**

In an alternate process, the chromatogram is stopped before any of the mixture to be separated has emerged from the column. The solvent is allowed to flow off, leaving a moist adsorbent which is extruded or pushed out of the column. The areas of the adsorbent containing the mixture components are separated and the substances are isolated by eluting them with a polar solvent.

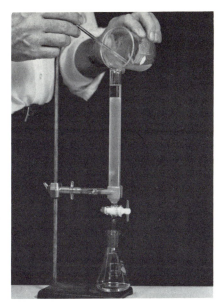

FIG. 1.6   FIG. 1.7

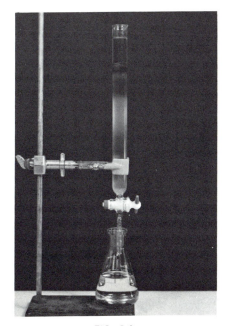
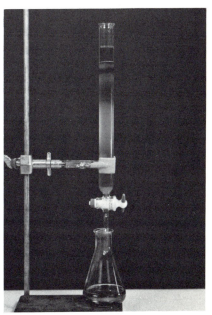

FIG. 1.8   FIG. 1.9

FIGS. 1.6 to 1.9. A slurry of silica gel in benzene is poured into a glass column (Fig. 1.6) and the solvent is allowed to flow out until the level is just over the adsorbent. A small circle of filter paper is placed on top of the column, and the dye mixture (see Fig. 1.1) is added with a dropper (Fig. 1.7). Additional benzene, as a developer, is allowed to flow through the column so that the colored bands are resolved (Fig. 1.8) and finally until one of them emerges from the bottom (Fig. 1.9). The time involved in this particular experiment was about one hour.

Gas Liquid Chromatography

In GLC the liquid stationary phase is present as a thin coating on a finely divided, inert solid support which is, in turn, packed in a small-diameter tube (1/16–1/4 in.) of moderate length (4–15 ft). The moving phase, an inert gas such as helium, is allowed to flow through the column, which is generally placed in an oven so that it can be heated to facilitate the separation of high-boiling materials. The sample mixture is **injected**, generally with a microsyringe, through a rubber septum into the column, and the components of the mixture flow through the column at varying rates and are detected as they emerge by a **detector** which transmits a signal to a recorder.

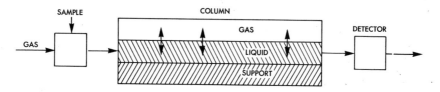

FIG. 1.10. A schematic GLC apparatus. The sample is injected into the gas stream and allowed to flow into the column where it can be partitioned between the gas phase and the liquid-coated support. The sample is separated and emerges from the column into the detector. (Redrawn from S. Dal Nogare and R. S. Juvet, Jr., "Gas-Liquid Chromatography," Interscience, New York, 1966, p. 24, with the permission of the authors and the publisher.)

Figure 1.10 schematically describes a GLC apparatus which consists of a few basic parts: a high-pressure carrier gas supply with a pressure regulator, a sample inlet system, a packed column of coated support, a detector, and a recording device for the detector. Figure 1.11 illustrates what occurs when a two-component liquid mixture is injected into the column.

The behavior of a specific compound on a particular column at a certain temperature and gas flow will be quite characteristic; thus the appearance of the compound at the detector will occur after a consistent length of time, the **retention time.** This retention time permits the qualitative identification of a compound in a mixture of others and, depending on the recording device, its quantitative estimation. Identification can also be carried out on a component which has been collected as it leaves the column.

1.3. APPLICATIONS

Qualitative Applications

First of all, qualitative applications of chromatography reveal the presence or absence of a specific compound in a sample. This is generally done by comparing the pure compound with the mixture on a single chromatogram (TLC) or by comparing the retention times (GLC).

Secondly, qualitative chromatography gives information on the complexity of a mixture. The mixture is chromatographed under various conditions and the number of spots (TLC) or peaks (GLC) indicates the minimum number of components. If some of the spots or peaks contain more than one substance, the mixture may be much more complex than it appears to be. Conversely, the purity of a compound can be studied. In this case, the compound is chromatographed *under several conditions* and concentrations. The presence of one spot (TLC) or one peak (GLC) under several conditions is a good criterion of purity and is frequently cited as such.

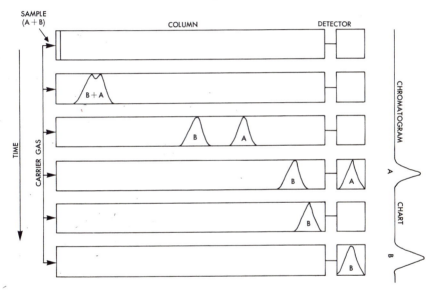

FIG. 1.11. A more detailed schematic diagram of a gas chromatogram. Solutes A and B can be seen to separate as they pass through the column and are recorded on a chart as they emerge from the column. (Redrawn from S. Dal Nogare and R. S. Juvet, Jr., "Gas-Liquid Chromatography," Interscience, New York, 1966, p. 15, with the permission of the authors and the publisher.)

Finally, qualitative chromatography can be used to establish a component or "fingerprint" pattern for a given tissue or for a given crude chemical or drug. Samples to be investigated can then be compared with this pattern to learn something about abnormalities of tissues or about adulterants of chemicals.

The techniques of choice for qualitative chromatography are thin-layer and gas chromatography. Column chromatography is too slow and laborious.

Of the two techniques, TLC is experimentally the simpler and requires less complex equipment. However, it is unsuitable for use with volatile compounds and, in many cases, does not have the resolving power of GLC. In general, GLC is used for the investigation of liquids and volatile materials (b.p. up to 400°C), and TLC is used where GLC equipment is not available or when the compounds are solids or are slightly volatile. In many cases, however, volatile materials are converted to nonvolatile derivatives for separation by TLC, and conversely, nonvolatile materials are frequently converted to volatile derivatives for separation by GLC. The great advantage of GLC, as shown later, is that quantitative information can be simultaneously obtained.

A major advantage of chromatography as a qualitative technique is the fact that only small amounts of materials are necessary. The lower limits are governed only by the detection techniques used and are generally less than 1 microgram.

Different types of scientists will seek qualitative information of various types. The synthetic chemist, organic or inorganic, will examine reaction mixtures to determine which reaction conditions give the cleanest products and which reactions do not take place (yielding only starting material). Furthermore, he might take samples from a reaction at timed intervals for chromatography. This type of experiment reveals information about the number of intermediates and the optimum reaction time (Figure 1.12). After this he may separate the reaction mixture into various fractions by extraction, distillation, or crystallization and inspect each of these fractions to learn where the product is. If the product is eventually isolated by column chromatography, the various fractions coming from the column are best examined by a qualitative method.

The natural products scientist, chemist or pharmacist, can monitor all of the steps in a complex separation scheme down to the final purification. The examination of tissues for the presence of various poisons or drugs is an important aspect of forensic chemistry. The use of "fingerprint" patterns as mentioned previously is extensive in industry as quality control and has been used in medicine to learn the fate of drugs after administration and to determine the occurrence of various metabolites in disease.

Quantitative Applications

Quantitative chromatography reveals how much of a mixture component is present, either as an absolute amount or relative to another component. There is little question that the method of choice for this work is gas chromatography, at least when the materials are volatile or easily convertible to volatile materials. Integration of the areas under the peaks of a recorder chart from a GLC chromatogram gives more accurate results with less labor

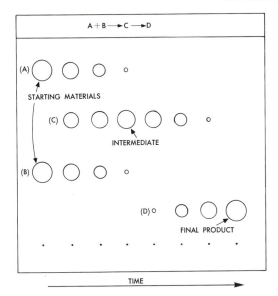

FIG. 1.12. A crude kinetic study by TLC of the reaction between A and B to yield the intermediate C which is, in time, converted to the final product D.

than TLC or column chromatography. Furthermore, GLC can easily be converted into a completely automated analysis system for industrial uses. TLC and column chromatography have, however, been used to obtain quantitative data.

Quantitative methods are used primarily for routine assay of samples, frequently in quality control of starting materials, intermediates and final products in an industrial process. Clinical chemistry is primarily a matter of quantitative assay for various materials in body tissues.

The organic chemist frequently uses quantitative methods for the establishment of product distributions or isomer ratios in the study of reaction mechanisms. In fact, the development of GLC during the last ten years has all but revolutionized this area. The special advantage is that the reaction mixtures can be assayed directly without any prefractionations which might alter the results.

Preparative Applications

Preparative methods are used to obtain reasonable quantities of mixture components in a pure state so that they can be characterized or used in further reactions. Column chromatography was developed primarily for this purpose and is still the method of choice. It should be noted, however, that both preparative GLC and preparative TLC are becoming quite important and are, in fact, preferred by many investigators.

In a discussion of preparative chromatography, one is forced to consider the major disadvantage of chromatography. This is the relatively small amounts that can be separated with what seems to be a great deal of labor. The amounts which can be separated on a single small column rarely exceed 10 g and are often smaller. The amounts which can be conveniently separated by preparative TLC are generally about 1 g. This is a major reason why many chemists avoid chromatography completely. The answer to this is fairly simple. One can spend days trying to find just the optimum solvent system for fractional crystallization and still not find it. It is generally better to invest a longer initial time in chromatography and be relatively certain that a clean separation will result.

Preparative separations are made in all fields of chemistry and biology, the recovery of insecticides from treated plants and animal tissues being a recent application. Chromatography is routinely used by organic chemists for product isolation, and any type of natural product work invariably involves chromatography in the final isolation steps.

1.4. THEORETICAL CONCEPTS

The major goal of this text will be to present a practical guide for the laboratory practice of chromatography. However, it might be worthwhile to present some of the theoretical concepts and terms involved, if only for the sake of pedagogy. Several approaches have been taken to the theory of the various techniques of chromatography. We have chosen the approach given by Johnson [1] as probably the simplest and easiest to understand. We will present the underlying ideas and the resulting equations *without derivation*. The derivations can be found in the original reference.

In the introduction to this chapter, we described the various partitions or distributions which can take place when a solute is introduced into a two-phase system such as a liquid-liquid system (partition), a liquid-solid system (adsorption), or a liquid-gas system (GLC). We then stated, rather blithely, that chromatography resulted when one of the phases was moved, mechanically, with respect to the other.

The actual conversion of a static situation into a flowing system can be visualized as the placing of a *number of distribution systems in series with one another*. Thus, the moving phase is allowed to equilibrate with one portion of stationary phase and then allowed to flow on to the next portion, and so on. Such a system is illustrated schematically in the left-hand portion

[1] M. J. Johnson in W. W. Umbreit, R. H. Burris, and J. F. Stauffer, "Manometric Techniques," Burgess Publishing Co., Minneapolis, 1964, p. 233.

of Figure 1.13. In this figure, the two phases are liquids and the same volume of each liquid is present in each distribution or equilibration stage. The lower phase in the diagram is the moving phase, but this is only a matter of mechanical design.

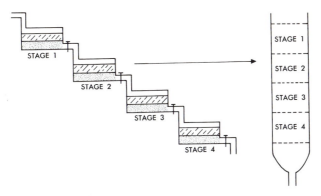

FIG. 1.13. A schematic approach showing the conversion of an intermittent partition system into a continuous column chromatogram.

The rate at which a solute will move in such a system will depend upon its relative solubilities in the two phases. This is expressed as a **partition coefficient** (K) and is defined as the concentration of solute present in one phase divided by the concentration of solute in the second phase after complete equilibrium has been achieved. The coefficient is constant for any given solute in any given system at a given temperature. Furthermore, it remains constant as the total amount of solute is increased to a point where one of the phases is saturated. By definition, in chromatographic systems, the numerator is the solute concentration in the moving phase and the denominator is the solute concentration in the stationary phase. The following equation can be used to express this relationship where [S] is the solute concentration. This coefficient can

$$K = \frac{[S]\ \text{mov}}{[S]\ \text{stat}} \tag{1.1}$$

be experimentally determined in a separatory funnel if solute is allowed to distribute between the phases and if the concentration in each phase is measured.

Suppose that we first consider the behavior, in a chromatographic system, of a solute that has a partition coefficient of 1. This will mean that the solute will distribute itself equally between two phases if the phases have the same volumes (as in Figure 1.13). Suppose further that we introduce

64 mg (64 makes the arithmetic easier) of solute into the first equilibration or stage 1. After equilibration, 32 mg will be in each phase (step 1 in Figure 1.14). Now open the valve and allow the moving phase to flow to a second stage which contains only stationary phase. Simultaneously, add a fresh portion of moving phase to stage 1. After equilibration, the amounts of solute will be distributed as shown in step 2 of Figure 1.14. If the process is repeated through five steps, the solute can be seen to be distributed with a maximum concentration in the middle stage. This data is plotted in Figure 1.15 (– – – –). If the system is extended to nine equilibra-

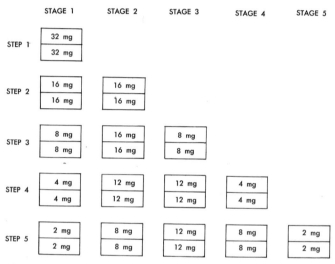

FIG. 1.14. Solute distributions which will result after various numbers of equilibrations.

tions and then to seventeen equilibrations, two additional curves will result (. . . . . . for nine and ——— for seventeen). Three important conclusions can be drawn from this experiment. A substance with a partition coefficient of 1, and thus equally distributed between the phases, will be found concentrated at the middle of the system. Second, if the number of equilibrations is increased, the solute will be concentrated in a smaller fraction of the total system. Finally, the distribution curve approximates a Gaussian distribution.

Now, suppose that a mixture of substances, a and b, is introduced into the system rather than a single pure solute. Suppose further that 64 mg of each component is used and that a and b have distribution coefficients of 3 and 0.33 respectively. The system can be operated as described above and the calculations can be made for each component without regard to the other. The result, after nine equilibrations, is shown in Figure 1.16. Stages 1–4 contain practically pure b and stages 7–9 contain practically pure a. If the total number of equilibrations had been increased, the peaks would have been sharper and the separations would have been more complete.

The system described thus far is not a true chromatographic process, but it tends to approach one. In a true chromatographic process, the equilibration stages are placed on top of one another and the valves are

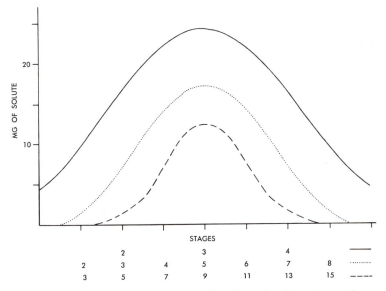

FIG. 1.15. The distribution of a single solute after five, nine, and seventeen equilibrations.

removed as in Figure 1.13. The system is normally full of stationary phase before the solute is introduced. During the development of the chromatogram, moving phase is introduced continuously at the top and removed from the bottom. An apparatus has been designed, however, that is exactly analogous to the described model, complete with discrete equilibrations and transfers. This is the Craig countercurrent apparatus, but a thorough discussion of this device is beyond the scope of this book.

Even though chromatography is continuous, it can, for theoretical purposes, be considered to consist of a series of equilibrations. These theoretical equilibrations are called **theoretical plates**, a term borrowed from distillation theory. If a column or thin layer consists of such a stack of theoretical plates, each plate can be considered to occupy a portion of column or layer length. The length is called the **HETP** or the **h**eight equivalent to a theoretical **p**late. The HETP is a measure of the efficiency of a chromatographic column and is about 0.002 cm for a good partition column.

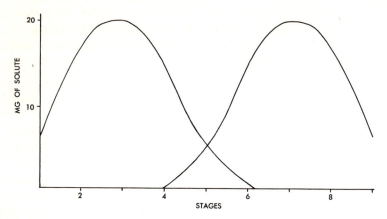

FIG. 1.16. The distribution of two solutes, a and b, with partition coefficients of 3 and 0.33, after nine equilibrations.

The partition coefficient must be modified or corrected before it can be applied in chromatographic processes. In the model described above, the volumes of the two phases were equal in each stage and in the entire process. In a column or TLC partition chromatogram, the stationary phase is generally present to a much lesser extent than is the moving phase. This correction can be made by multiplying the partition coefficient by the volume of the moving phase divided by the volume of the stationary phase. The new term will be designated B and it is defined in Equation (1.2).

$$B = K \times \frac{\text{volume of moving phase}}{\text{volume of stationary phase}} \tag{1.2}$$

The volumes of the phases refer only to the volumes present in the column, and these can be obtained by measuring the amounts used in the column preparation. From a knowledge of the B values of solutes, it is possible to predict the location, on a column or TLC layer, of the various peaks.

In order to do this, however, we must have some unit of column measurement which can be used to denote peak location. This is the R_f unit. The R_f is actually defined as the rate of movement of a given solute (measured at its maximum) divided by the rate of movement of the moving phase. In the case of a TLC layer, both solute and solvent move for the same number of minutes and the R_f can be expressed in length units (Figure 1.5).

In the liquid-liquid intermittent system described above, the R_f is equal to the number of the equilibration stage containing the maximum solute divided by the total number of equilibrations. This is true, however, *only because just enough moving phase* has been introduced to fill the system. To put it another way, the moving phase originally present in the first stage has moved only to the last stage, but has not left the system. The R_f of the solute in Figure 1.15 is 0.5 (4 ÷ 8). Since the volumes of the two phases were, by definition, the same, B is equal to K. The R_f and B are related by Equation (1.3).

$$\frac{B}{B+1} = R_f \tag{1.3}$$

The volume of moving phase necessary to fill the system, or to displace all of the phase present in the system before the chromatography was begun, is called the **holdup volume.** This term will be designated as v. It can be measured experimentally in several ways. The best way is to place a marker dye in the solute or solute mixture which has no solubility in the stationary phase. The amount of solvent necessary to move the dye out of the bottom of the column is then the holdup volume. A second way would be to measure the amount of moving phase used to pack the column so that the solvent level is at the top of the stationary phase. The R_f of a solute on a column is then equal to the distance of the peak (or band) from the top of the column divided by the length of the column, after one holdup volume of moving phase has been passed through.

In the discussion thus far, we have been concerned only with the separations which take place during a development with one holdup volume of moving phase. This is, of course, the situation in TLC and in the type of column chromatography in which the adsorbent is extruded and sectioned to obtain the purified solutes (see p. 99). However, in GLC and in most column chromatography, the solute bands are moved off of the column or are eluted and either measured or isolated. This type of chromatogram is called a **flowing chromatogram** or an **elution chromatogram.** In this case fresh moving phase is introduced well beyond one holdup volume, and the peaks as shown in Figure 1.15 are washed off of the stationary phase. It then becomes desirable to know how much moving phase, V, will be required to move a given peak off of a column. This can be calculated using Equation (1.4), which relates B values, R_f values, and the holdup volume.

$$V = \left(\frac{B+1}{B}\right) v = \frac{v}{R_f} \tag{1.4}$$

The number of theoretical plates in a column can be approximated for a flowing chromatogram by Equation (1.5). In the equation, n is the number of plates, V is the volume of moving phase needed to remove a solute peak from the chromatogram, and v is the holdup volume. The term w is obtained from the contour of the solute peak as it comes off of the column.

$$n = \frac{8V(V - v)}{w^2} \tag{1.5}$$

Such a contour is obtained by collecting the column effluent in fractions and analyzing the fractions. The amounts of solute are then plotted against the effluent volume. The term w is measured (in volume units as are V and v) at a distance 36.8% of the way from the base line to the peak. If the number of theoretical plates in a column, the holdup volume of a column, the volumes of the two phases, and the partition coefficients of two or more solutes are all known, it is possible to calculate the volume of solvent, V, needed to elute each peak and the value of w for each band which will give some idea of the extent of peak overlap.

Suppose that compounds a and b as given above with partition coefficients of 3 and 0.33 were separated on a column 10 cm high which had 500 theoretical plates, a holdup volume of 10 ml, and a ratio of moving phase to stationary phase of 10:1. The B value of a is then 30 (Equation (1.2)) and the volume needed to elute it from the column will be $\frac{30 + 1}{30}$ × 10 or 10.3 ml (Equation (1.4)). The w value would be

$$\sqrt{\frac{8 \times 10.3(10.3 - 10)}{500}} = \sqrt{\frac{82.4 \times 0.3}{500}} = 0.7 \text{ ml}$$

(Equation (1.5) rearranged). Thus the band of a would come off of the column very quickly in a very narrow zone, most of which was present in 0.7 ml of effluent. A similar calculation for b would show that it would be eluted by 13 ml of moving phase with a w value of 0.8 ml. The band of a will be fairly well separated from b since the solvent emerging between the ranges covered by the two w values is

$$\left(13 - \frac{0.8}{2}\right) - \left(10.3 + \frac{0.7}{2}\right) = 12.60 - 10.75 \text{ or } 1.85 \text{ ml}$$

The separation could be improved in several ways. If a moving phase were chosen so that the solutes were less soluble and the partition coefficients were lowered, the volumes of moving phase needed to move the solutes out of the column would be more different. A longer column would increase the holdup volume and the number of theoretical plates, thus enhancing the differences between the V values and compressing the bands as they were eluted.

This theoretical approach rests upon some rather tenuous assumptions and must be considered as only an approximation of the true picture. First, a complete equilibrium at each theoretical plate is assumed. This could be true only if the moving phase were moved infinitely slowly through the column. Second, diffusion of solute takes place within the column which tends to spread the solute over large areas of the column than are predicted. This is actually a function of time, but it has been completely neglected. Third, the partition coefficients have been assumed to be constant over an infinite range. This is true only to the extent of solubility of the solute in the phase present in the smallest amount. Finally, the conclusions reached are valid only in a liquid-liquid system. A much more thorough treatment can be found in Chapters 3 and 6 of "Chromatography," edited by Heftmann and cited in the Appendix. (Chapter 3 was written by J. C. Giddings and Chapter 6 was written by R. A. Keller and J. C. Giddings.) Still another treatment can be found in the text of Laitinen.[2]

Adsorption Systems

Adsorption chromatography is much more complex than partition chromatography. This is primarily because the distribution coefficients between solids and liquids are not constant as are the partition coefficients. This can be seen in Figure 1.17 where the relative amounts of solute in the two phases are plotted as the total amount of solute per given amount of the phases is increased. In a liquid-liquid system, the relative amount of solute in each phase remains constant and a straight line results (a). In the case of a solid-liquid system, the relationship is generally best expressed as

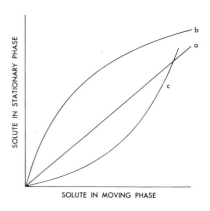

FIG. 1.17. A plot of the amounts of solute in the moving phase vs. the amounts of solute in the stationary phase for a typical partition system (a) and two adsorption systems (b) and (c).

SOLUTE IN STATIONARY PHASE

SOLUTE IN MOVING PHASE

[2] H. A. Laitinen, "Chemical Analysis," McGraw-Hill, Inc., New York, 1960, p. 492.

a convex curve (b) or less frequently, as a concave curve (c). The shapes of these curves, which are called sorption isotherms, are a property of the given solutes in the given systems and cannot be modified. They can be measured experimentally by adding various amounts of solute to a given amount of the stationary and moving phases and measuring the concentrations after equilibration. This means that the B values of the solutes would depend upon how much solute was present in any given area. Most of the theoretical work done in adsorption systems has been done with very small amounts of solute so that one is working in the more linear areas of curves (b) and (c).

The more common convex curve is associated with **tailing** on a thin layer or column while the less common concave curve is associated with **bearding.** In tailing, portions of solute lag behind the major concentration, and in bearding, portions precede the major concentration. These effects can be reduced by using smaller amounts of solute or larger columns.

A theoretical treatment of adsorption chromatography can be found in Chapter 4 of the book "Chromatography," edited by Heftmann and cited in the Appendix. (The chapter was written by L. R. Snyder.)

Thin-Layer Chromatography

The theory of TLC is complicated by several factors, at least as it can be compared with the model previously described. The rate of solvent flow is not constant, but varies as the solvent front advances through the layer. Furthermore, the rate depends upon the liquid used and is beyond control. The solvent is constantly moving into areas of dry adsorbent which tends to involve heats of wetting and distorted equilibria. The spots can undergo a lateral diffusion and are probably not uniformly deposited on the adsorbent to begin with. However, some approaches have recently been made to the theory of TLC by Stewart,[3] Snyder,[4] and Thoma and Perisho.[5] Probably the most important conclusion drawn thus far is that the greatest degree of resolution in a thin layer occurs in the area around an R_f of 0.3–0.4.

Gas-Liquid Chromatography

The theory of GLC is complicated by the fact that one of the phases, the gas phase, is compressible, and that it is at different pressures from one end of the column to the other. However, it has probably been more intensively studied than any of the other types of chromatography discussed in this book. Excellent discussions can be found in the book "Gas-Liquid Chromatography" by Dal Nogare and Juvet and in Chapter 6 of "Chromatography," edited by Heftmann. Both books are cited in the Appendix.

[3] G. H. Stewart, *Separation Science* **1,** 747 (1966).
[4] L. R. Snyder, *Anal. Chem.* **39,** 698 (1967).
[5] J. A. Thoma and C. R. Perisho, *Anal. Chem.* **39,** 745 (1967).

2

Thin-Layer and Column Chromatography—Choice of a System

2.1. INTRODUCTION

The most difficult problem in chromatography is the choice of a system that will perform the desired separation. Should one use an adsorption system or a partition system? What adsorbent or support should be used? Which solvents or mixtures of solvents? When should gas chromatography be used? Partial answers to these questions can be found in the literature, in experience, and in an understanding of some of the basic phenomena involved. Complete answers do not exist. Successful separations can be carried out only by careful experimentation preceded by some shrewd planning.

In a discussion of this nature, TLC and column chromatography can be considered closely related. Both involve the same types of chromatography: adsorption and partition. The major difference lies in the location of the stationary phase. In TLC it is in a thin layer on a glass plate, and in column chromatography it is in a column held in a glass tube. In fact, TLC has often been called "open-column chromatography."

Gas chromatography is inherently different in that a volatility phenomenon is involved. The advantages and disadvantages have been discussed on p. 106 and the precise choice of a GLC system will be a subject of Chapter 5. There is little question that it is the technique of choice for working with all volatile compounds.

The remainder of this chapter will be devoted to a discussion of the term "polarity," the choice between adsorption and partition systems, the basic phenomena underlying these methods, the manipulation of these systems, and finally, some ideas on the choice of precise chromatographic conditions.

2.2. POLARITY

The term polarity is a much used and often overused term in chromatography. Basically, it means the possession by a molecule of separate positive and negative centers arising from the atoms involved and their arrangement or configuration. Thus, the molecule may have a dipole moment and may be attracted to other molecules which also possess polarity. The extent of separation between the centers or poles will determine the degree of polarity and, hence, the degree of attraction. In chromatography, however, the meaning is broadened to include such properties as hydrogen bonding and polarization phenomena. It is essentially a relative term and is applied to solvents, solutes, and adsorbents. Water has a strong permanent dipole by virtue of its molecular geometry and electron configuration and is hydrogen-bonded. It is considered a very polar solvent. The oxygenated organic compounds such as alcohols, ketones, esters, and ethers have less strong dipoles, are less hydrogen-bonded and are therefore less polar than water. The hydrocarbons are the least polar of all. Although benzene has a symmetrical structure and no appreciable dipole, it does have an electron cloud that can be displaced or polarized by the approach of a polar material. Thus it behaves as a more polar substance than the aliphatic hydrocarbons. In fact, it can be safely stated that aromatic compounds of all types are more polar than their aliphatic counterparts. The relative polarity of solvents is manifested in their **dielectric constants,** which can be measured directly.

Adsorbents can be polar or nonpolar. Silica gel and alumina are both polar adsorbents, and will, in fact, be the only adsorbents discussed in this text. These adsorbents will adsorb polar solutes more tightly than they will less polar solutes. Thus, alcohols would be adsorbed more tightly than ethers, which would be adsorbed more tightly than hydrocarbons, etc. There are also some nonpolar adsorbents, the chief one being charcoal.

The term polarity also has a bearing on the solution phenomena which underlie both adsorption chromatography (in the liquid phase) and partition chromatography (in both liquid phases). Here, essentially, one can say that "like dissolves like"; polar solvents tend to dissolve polar solutes more readily than they do nonpolar solutes and vice versa.

The relative polarities of some common solvents are given in Table 2.1, in which the more polar solvents are at the bottom. This is the so-called eluotropic series which describes the polarity as it is manifested in chromatographic systems. In general, the polarity of organic compounds *increases with the number of functional groups and decreases with increasing molecular weight.* Thus, the simple carbohydrates are considered quite polar substances by virtue of their many hydroxyl groups. Within an homologous series, the polarity of the higher members is less than that of the lower ones because the more polar functional group becomes relatively less important in the larger molecules.

Polarity is even used to describe the properties of molecules containing actual positive and negative charges. Thus, salts of acids and bases and the amino acids (in their zwitterionic forms) are considered very polar substances indeed.

TABLE 2.1

Eluotropic Series of Solvents [a]

Light petroleum (pet. ether, hexane, heptane, etc.)
Cyclohexane
Carbon tetrachloride
Trichloroethylene
Toluene
Benzene
Dichloromethane
Chloroform
Ethyl ether
Ethyl acetate
Acetone
- n-Propanol
Ethanol
Methanol
Water

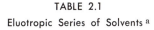 more polar

[a] W. Trappe, *Biochem. Z.* **305**, 150 (1940).

2.3. ADSORPTION VS. PARTITION

Both adsorption and partition separations depend primarily upon the differences in polarity between the solutes being separated, since polarity is the major factor governing both solubility and adsorption (at least on the polar adsorbents considered in this text). The two phenomena differ, however, in some of their secondary features.

Partition processes are dependent upon solubility in two liquids and are quite sensitive to small differences in molecular weight between solutes. For this reason, the members of an homologous series are generally best separated by partition chromatography, especially those members containing more than four or five carbons.

Adsorption processes, on the other hand, are quite sensitive to steric or spatial differences between solutes. The extent to which the solute can be accommodated on an adsorbent surface will depend, in part, upon its configuration, and this, in turn, will determine its adsorption relative to other solutes. This relative adsorption determines the separability. Thus, adsorption chromatography would be the method of choice for the separation of similar molecules having a slightly different stereochemistry.

The choice between adsorption and partition chromatography will be governed by three factors: the relative experimental difficulty of the two methods; the purpose and goal of the separation; and primarily, by the type of compounds being separated.

Adsorption chromatography is experimentally easier than partition chromatography. As a first approximation in adsorption chromatography, the adsorbent is constant and the solvents are varied over the eluotropic series in search of a polarity that will effect the separation. In partition chromatography, neither of the two phases can be maintained at constant polarity and it is difficult to carry out systematic experimentation. It is quite reasonable, then, to try adsorption first and partition second.

Adsorption chromatography can be used for the separation of larger quantities than can partition chromatography. Thus, it is a better preparative method. Adsorption chromatography, particularly as practiced in columns, can be used for the separation of mixtures of solutes which vary widely in their polarity and structure. Partition chromatography, in contrast, is best applied to mixtures of fairly similar solutes. It is quite common to carry out preliminary large-scale fractionations of very crude mixtures by adsorption processes and to submit these crude fractions to partition chromatography for final separation. Partition chromatography has a higher resolving power than adsorption and is, under proper circumstances, a better qualitative method.

It would seem reasonable in light of the previous discussion to assume that any mixture can be resolved by adsorption chromatography if one can find a solvent of the correct polarity. Sad to say, this is not quite true. When solutes are polar and tightly adsorbed to a surface, it is necessary to use highly polar solvents to move them. Frequently, these highly polar solvents will tend to overwhelm small differences between solutes and to produce no separations. As a general rule, highly polar materials such as carbohydrates, amino acids, and nucleotides are separated by partition methods, and relatively less polar materials such as hydrocarbons and monofunctional organic molecules are separated by adsorption methods. Keeping in mind that polarity increases with the number of functional groups and decreases with increasing molecular weight or carbon content, it is possible to assemble a diagram such as shown in Table 2.2 which might help in the choice of a system. The polarity increases from left to right, primarily due to increased dipoles in the substances involved, and from top to bottom due to increased number of functional groups. A slight decreasing tendency from bottom to top is due to increased carbon chains. Thus, the most polar substances (those most amenable to partition separations) will be found in the lower right-hand corner of the diagram and the least polar (most amenable to adsorption separations) will be found in the upper left-hand

TABLE 2.2

Adsorption vs. Partition

Polarity of Functional Groups [a]

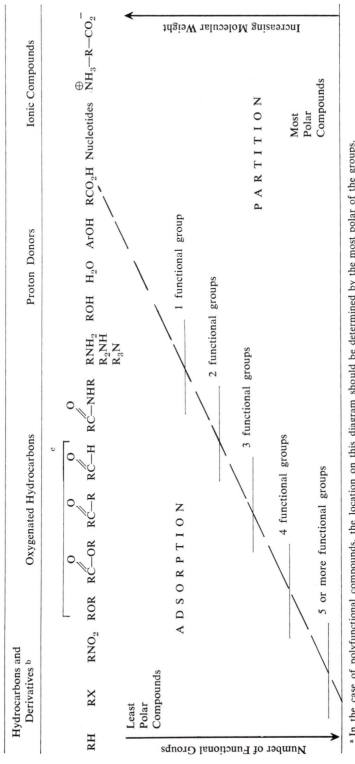

a In the case of polyfunctional compounds, the location on this diagram should be determined by the most polar of the groups.
b Double bonds should be considered functional groups although it should be noted that a double bond does not increase polarity as much as the functions. Aromatic rings also increase polarity, but more so than three double bonds.
c The esters, ketones, and aldehydes have similar polarities.

25

corner. A diagonal line from upper right to lower left divides the diagram into those compounds best separated by partition and those best separated by adsorption. Note that the dividing line begins (at the bottom) to the right of the hydrocarbons and ends (at the top) to the left of the amino acids and nucleotides. This would mean that hydrocarbons are always separated by adsorption methods and that amino acids are always separated by partition methods. This is valid. Between these extremes along the dividing line is an area where either process might be applicable.

2.4. ADSORPTION PROCESSES

Nature

Two factors are involved in adsorption chromatography: the forces attracting solutes to adsorbents and the forces tending to remove them from the adsorbents so they will move with the moving phase and be separated.

The forces which cause adsorption of neutral solutes are primarily dipole-dipole attractions (between polar adsorbents and polar solutes), hydrogen bonding (between the hydroxyl groups on silica gel especially and various basic atoms such as oxygen and nitrogen on the solutes), and polarizability forces (between polar adsorbents and solutes such as aromatic materials that can be polarized, see p. 22). The order of adsorption to polar adsorbents [1] is shown in both Table 2.1 (low at the top and high at the bottom) and Table 2.2 (increasing from left to right and top to bottom).

The chromatography of acids and bases is somewhat different. Of the adsorbents discussed in detail in this book, one (silica gel) is relatively acidic and one (alumina) is relatively basic. If acids are chromatographed on alumina, they will be tightly bound to the adsorbent by ionic forces and quite difficult to move and resolve. The same is true of bases on silica gel. Thus, bases should be chromatographed on alumina, where they will be adsorbed to an extent similar to alcohols. Acids and phenols should be separated on silica gel, where they will be somewhat more tightly adsorbed than alcohols.

There are two forces which cause solutes to move with the solvent in a chromatogram. The first of these is the tendency to dissolve in and move with the solvent; a phenomenon often called **elution.** In this case, the ideal solvents should dissolve the solutes and should be just good enough, as solvents, to compete with the adsorptive power of the adsorbent. This type of situation probably prevails when nonprotonic solvents such as hydrocarbons, ethers, and carbonyl compounds are being used as developing solvents.

The second force tending to move solutes in a chromatographic system is a displacement phenomenon. The solvent molecules tend to compete

---

[1] In the case of nonpolar adsorbents such as charcoal, this order is reversed.

with the solute for adsorption sites and therefore move them along. This is sometimes called **displacement,** and the type of chromatography which is based upon this notion alone is called **displacement analysis.** In a sense, the elution described above is a pulling force and displacement is a pushing force. Elution phenomena probably prevail when nonprotonic solvents (hydrocarbons, ethers, esters, chlorocarbons, ketones) are used, but the introduction of alcohols as solvents almost surely produces some displacement.

The line of demarkation between these two phenomena is quite blurred. Most separations can be readily carried out and understood if it is assumed that only elution is taking place and that the protonic solvents are merely very polar solvents.

Manipulation

Adsorption processes can be manipulated by changing the nature of the adsorbents or by changing the nature of the solvents. Solvent manipulation is easier and is more generally used.

First of all, a major variation lies in the adsorbent chosen. Acidic materials should be separated on silica gel and basic materials should be separated on alumina. Neutral materials can be separated on either, although frequently one will work better than the other. Silica gel is more frequently used in TLC whereas alumina is more frequently used in columns.

After this preliminary choice, the properties of the adsorbents can be varied further. Alumina can be prepared in several activities (or polarities) and in forms that are more or less acidic. This is frequently done for column work and will be more carefully discussed later (p. 90). Silica gel is sometimes modified by the addition of acids or bases and especially by the addition of complexing agents such as 2,4,7-trinitrofluorenone or silver ion. These are more frequently techniques of TLC and will be discussed later (p. 56).

The solvents are manipulated primarily by varying and mixing them to produce an appropriate polarity for a given separation. The relative polarities of a series of solvents are given in the eluotropic series in Table 2.1. This is an ideal interpretation and is subject to some severe deviations. Certain solvents sometimes appear to have special properties for the separation of certain molecules. For example, one might prepare a blend of benzene and ether which would have exactly the same polarity as chloroform (see Table 2.1), and yet find that chloroform will produce a separation and the blend will not. Such matters require experimentation and experience.

It should be noted that the polarity resulting from blending solvents is not a linear function, but logarithmic. The first small amount of ether added to benzene changes the polarity a great deal whereas the difference between 40 and 50% ether is not so appreciable. Solvents having polarities fairly close together should be blended. In order to find a polarity similar to chloroform, for example, benzene and ether should be blended, rather than the extremes, hexane and methanol. The polarities produced by blending solvents are shown in the unique diagram in Figure 2.1. The data were measured by Neher and von Arx [2] and are based upon the R_f measurements of 20 steroids on silica gel. They should, however, be applicable to adsorption chromatography in general. The polarity increases from left to right in the figure and any vertical line (such as the dotted line) will connect mixtures of similar polarities.

Choice of a Specific System

The search for a specific system for the separation of a mixture is best carried out by TLC. The technique is such that a large number of adsorbent-solvent systems can be explored in a short time. A single column chromatogram is frequently a lengthy experiment. In theory, one should be able to transpose systems from TLC to column chromatography. In practice, this is possible, but not easy. This transposition will be specifically discussed in Chapter 4. At the very least however, TLC studies can yield general information about the polarity of a mixture.

An experimental approach to this problem can be outlined as follows. Prepare some thin layers of the adsorbent or adsorbents to be investigated, perhaps on microscope slides (see Chapter 3). Spot the mixture on four chromatograms and develop them in the four solvents underlined in Table 2.1, hexane, benzene, ether, and methanol. Visualize the chromatograms (see p. 47) and compare the results. In this manner, one may find a pure solvent that will resolve the mixture, or at least learn something of its general properties. Frequently, it is possible to "bracket" the properties of the mixture, that is, to find one solvent which will move the spots too far and one which will not move them far enough. Mixtures of these two solvents can then be investigated as well as the other pure solvents which lie between them in Table 2.1. In general, the best separations are obtained in the lower half or two-thirds of the developed layer (see p. 20).

There are also certain solvent additives that will reduce streaking and give improved separations. When acidic materials are being chromatographed on silica gel, a drop of acetic acid (or about 1%) may be added to the developing solvent. Conversely, when basic materials are being chromatographed, a drop of ammonium hydroxide or diethylamine may be added to the solvent system. This is not necessary for the chromatography of bases on alumina. These acid and base additives help to buffer the materials being separated so that they remain completely in a nonionic form and give more compact spots.

[2] R. Neher, "Steroid Chromatography," American Elsevier Publishing Company, Inc., New York, 1964, p. 249.

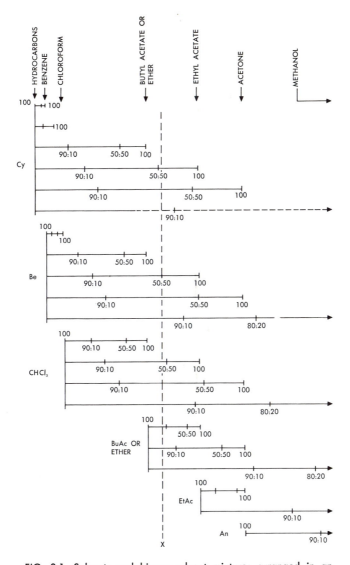

FIG. 2.1. Solvents and binary solvent mixtures arranged in an "equi-eluotropic" series on the basis of average R_f values for 20 steroids. Any vertical line such as the dotted line X will connect solvent mixtures having an equal ability to move solutes in an adsorption system. Pure methanol is somewhere out of the figure to the right. Solvent abbreviations are: Cy, cyclohexane; Be, benzene; BuAc, butyl acetate, EtOAc, ethylacetate; An, acetone. (Reproduced from R. Neher, "Steroid Chromatography," [2] through the courtesy of the author and Elsevier Publishing Co.)

29

If the system is to be used on a column, perhaps a gradient elution (p. 96) between the two solvents which "bracket" the mixture can be used. In the separation of very complex mixtures such as those mentioned on p. 95, gradient elution covering the entire eluotropic series may be required.

Some applications of adsorption chromatography are given in Table 2.3. These are taken from the book by Bobbitt on TLC.[3] They should, however, give some idea of the general conditions which would give satisfactory separations on adsorption columns as well as in TLC. The visualization reagents are those used in TLC. Additional suggested reagents are placed in parentheses. The solvent abbreviations are as follows: EtOEt—ethyl ether; EtOAc—ethyl acetate; MeOH—methanol; EtOH—ethanol; and BuOH—n-butanol. The phrase "U.V. on phosphors" means U.V. light on phosphor-containing adsorbents, see p. 42. The "G" designation on the adsorbent means that it contains a plaster-of-Paris binder. The absence or presence of this binder should make little difference in an adsorption process.

2.5. PARTITION PROCESSES

Nature

In its basic form, a partition system and the forces that govern its operation are much simpler than an adsorption system. It is made up of two liquids that are only partially miscible in one another. Because of this immiscibility requirement, one of the liquids will tend to be much more polar than the other. Either of the liquids may be a pure solvent or a mixture of considerable complexity.

The factors that will dictate whether a solute will remain in the stationary phase or move with the moving phase will be its solubilities in the two liquids. The general aspects of solubility have been considered previously on p. 22 in respect to the moving phase in adsorption chromatography. The extent to which a solute distributes between two liquids can be measured in a separatory funnel. Such measurements produce constants known as **partition coefficients.** Basically, materials that have different coefficients can be separated. Solutes which are more soluble in the moving phase will move faster than those which are less soluble. As a corollary, solutes which are more soluble in the stationary phase will tend to move more slowly than those which are less soluble.

The problems involved in placing one of the liquid phases on an appropriate support and holding it stationary during chromatography are mechanical problems and will be discussed in the sections devoted to techniques.

[3] J. M. Bobbitt, "Thin-Layer Chromatography," Reinhold Book Corporation, New York, 1963, pp. 128–182. Original literature is cited therein.

TABLE 2.3

Suggested Systems for Adsorption Chromatography

Compounds	Adsorbent	Developer	Visualization [a]
1. Long-chain (over eight carbons) aliphatic ketones	silica gel G	a. benzene–EtOEt mixtures b. toluene–EtOEt mixtures c. pet. ether–EtOEt mixtures	phosphomolybdic acid (2,4-DNPH)
2. Aliphatic and aromatic aldehydes and ketones	aluminum oxide	a. benzene b. benzene–EtOH (98:2) (95:5) (90:10) c. $CHCl_3$ d. EtOEt e. pet. ether–benzene (1:1)	2,4-DNPH
3. Aromatic aldehydes and ketones	silica gel G	hexane–EtOAc (4:1) (3:2)	(U.V. on phosphors)
4. Vanillin and other substituted benzaldehydes	silica gel G	a. pet. ether–EtOAc (2:1) b. hexane–EtOAc (5:2) c. $CHCl_3$–EtOAc (98:2) d. decalin–CH_2Cl_2–MeOH (5:4:1)	(U.V. on phosphors)
5. 2,4-Dinitro-phenylhydrazones of aldehydes	silica gel G	a. benzene–pet. ether (3:1) for aliphatic b. benzene–EtOAc (95:5) for aromatic	colored
6. Miscellaneous alkaloids	silica gel G	a. $CHCl_3$–acetone–diethylamine (5:4:1) b. $CHCl_3$–diethylamine (9:1) c. cyclohexane–$CHCl_3$–diethylamine (5:4:1) d. cyclohexane–diethylamine (9:1)	(Dragendorff's Reagent)

[a] Directions for the preparation of these reagents are found on p. 69.

TABLE 2.3 (cont.)

Compounds	Adsorbent	Developer	Visualization [a]
	aluminum oxide	e. benzene–EtOAc–diethylamine (7:2:1) f. $CHCl_3$ g. cyclohexane–$CHCl_3$ (3:7) plus 0.05% diethylamine	
7. Alkaloids and barbiturates in toxicology	silica gel G	a. MeOH b. $CHCl_3$–EtOEt (85:15)	Dragendorff's Reagent
8. Strong-base amines	unbound aluminum oxide	a. acetone–heptane (1:1) b. $CHCl_3/NH_3$ (sat. at 22°)–EtOH (96%) (30:1)	I_2 vapor-U.V. (Dragendorff's Reagent)
9. Sugar acetates and inositol acetates	silica gel G	benzene with 2–10% MeOH	H_2O
10. Aldose 2,4-DNPH's	a. aluminum oxide G b. silica gel G	a. toluene–EtOAc (1:1) b. toluene–EtOAc (3:1) (1:1)	a. colored compounds b. NaOH
11. Cardiac glycosides	silica gel G	a. $CHCl_3$–MeOH (9:1) b. $CHCl_3$–acetone (65:35)	H_2SO_4 and heat
12. Dicarboxylic acids	silica gel G	a. benzene–MeOH–HOAc (45:8:4) b. benzene–dioxane–HOAc (90:25:4)	bromphenol blue acidified with citric acid
13. p-Hydroxybenzoic acid esters	silica gel G	pentane–HOAc (88:12)	(U.V. on phosphors)
14. Sulfonamides	silica gel G	$CHCl_3$–EtOH–heptane (1:1:1)	p-dimethylamino-benzaldehyde, acidified (U.V. on phosphors)
15. Food dyes	silica gel G	a. $CHCl_3$–acetic anhydride (75:2) b. benzene	$SbCl_3$ in $CHCl_3$

TABLE 2.3 (cont.)

Compounds	Adsorbent	Developer	Visualization [a]
		c. methyl ethyl ketone–HOAc–MeOH (40:5:5)	
16. Misc. essential oils	silica gel G	benzene–CHCl$_3$ (1:1)	SbCl$_3$ in CHCl$_3$
17. Coumarins	silica gel G	a. pet. ether–EtOAc (2:1) b. hexane–EtOAc (5:2)	(U.V. on phosphors)
18. Alkali metals—Na^{1+}, Li^{1+}, K^{1+}, Mg^{2+}	purified silica gel G—see p. 48	EtOH–HOAc (100:1)	acid violet (1.5% aqueous soln.)
19. 24 ferrocene derivatives	silica gel G	a. benzene b. benzene–EtOH (30:1) (15:1) c. propylene glycol–MeOH (1:1) d. propylene glycol–chlorobenzene–MeOH (1:1:1)	(U.V. on phosphors)
20. Misc. lipids	silica gel G	a. EtOEt b. isopropyl ether c. isopropyl ether–HOAc (98.5:1.5)	(H$_2$SO$_4$ and heat)
21. Fatty acid methyl esters	silica gel G	hexane–EtOEt mixtures (up to 30% EtOEt)	a. I$_2$ b. H$_2$SO$_4$ and heat
22. Glycerides	silica gel G	a. Skellysolve F–EtOEt (70:30) b. (10:90) for monoglycerides c. (70:30) for diglycerides d. (90:10) for triglycerides e. (60:40) (35:65) (85:15)	H$_2$SO$_4$ and heat
23. Long-chain aliphatic alcohols	silica gel G	a. pet. ether–EtOEt–HOAc (90:10:1)	H$_2$SO$_4$ and heat

TABLE 2.3 (cont.)

Compounds	Adsorbent	Developer	Visualization [a]
		b. pet. ether– EtOEt (20:80) (10:90) (70:30)	
24. Phenols	a. starch-bound silicic acid b. starch-bound silicic acid kieselguhr (1:1)	a. xylene b. xylene–$CHCl_3$ (3:1) (1:1) (1:3) c. $CHCl_3$	(U.V. on phosphors)
25. Phenols and phenolic acids	starch-bound silicic acid with phosphors	a. Skellysolve B– EtOEt (3:7) b. Skellysolve B– EtOAc (1:3) c. Skellysolve B– acetone (3:1)	(U.V. on phosphors)
26. Phenols	plaster-of-Paris bound silica gel	a. hexane–EtOAc (4:1) (3:2) b. benzene–EtOEt (4:1) c. benzene	(U.V. on phosphors)
27. 3,5-Dinitroben- zoates of alcohols and phenols	silica gel G	a. benzene–pet. ether (1:1) b. hexane–EtOAc (85:15) (75:25) c. toluene–EtOAc (90:10)	colored com- pounds
28. Steroids	silica gel G	a. benzene b. benzene–EtOAc (9:1) (2:1) c. cyclohexane– EtOAc (9:1) (19:1) d. 1,2-dichloro- ethane	$SbCl_3$ in $CHCl_3$
29. 19-Nor-steroids	silica gel G	EtOAc–cyclo- hexane mixtures	$SbCl_3$ in $CHCl_3$
30. Blood cholesterol and cholesterol esters	silica gel G	a. benzene b. benzene–EtOAc (9:1) c. 1,2-dichloro- ethane d. $CHCl_3$	$SbCl_3$ in $CHCl_3$

TABLE 2.3 (cont.)

Compounds	Adsorbent	Developer	Visualization [a]
31. Triterpenoids	silica gel G	a. isopropyl ether–acetone (5:2) (19:1) b. isopropyl ether c. cyclohexane d. benzene e. CH_2Cl_2	(H_2SO_4 and heat)
32. Carotenes and fat-soluble vitamins A, D, E, K	unbound aluminum oxide	a. MeOH b. CCl_4 c. xylene	H_2SO_4 and heat
33. Thiophene derivatives	a. silica gel G b. aluminum oxide G	a. benzene–$CHCl_3$ (9:1) b. MeOH c. pet. ether	(U.V. on phosphors)
34. Plasticizers (phthalates, phosphates and other esters)	silica gel G with phosphors	a. isooctane–EtOAc (90:10) b. benzene–EtOAc (95:5) c. butyl ether–hexane (80:20)	a. U.V. on phosphors b. I_2

Manipulation

A partition system is manipulated by changing the nature of the two liquid phases, generally through the addition solvents or buffers. In these cases, the more polar liquid is generally water. The tendency of water to dissolve solutes can be changed by adding salts (to produce a salting-out effect), buffers (to dissolve or precipitate ionic materials such as amino acids), or complexing agents (borate ion for the complexing of glycols and sugars). The less polar phase of a partition system is generally an organic solvent such as *n*-butanol, benzyl alcohol, phenol, EtOAc, $CHCl_3$, or C_6H_6 with some solubility in the polar liquid (water). The ability of organic liquids to dissolve solutes can be varied by adding more or less polar solvents.

The problem is scarcely this simple, however, because the two phases must be in equilibrium with one another during the chromatography. Any material that is added to either phase will tend to distribute itself between the two phases and change the properties of *both*. For example, suppose that a water–benzene system is the basic starting point. In order to increase the power of the benzene to dissolve solutes, methanol is added. The methanol will not remain in the benzene, however, but will move primarily into the water, thereby changing the solvent power of both phases. This is in contrast with an adsorption system, where one can hold the adsorbent fairly constant and vary the solvents widely.

Choice of a Specific System

In adsorption chromatography, one can frequently find a system for the separation of a mixture by trial and error. In partition chromatography the systems are often complex and it may be wise to consult the literature before spending too much time in experimentation. A number of elementary systems are given in Table 2.4. Additional data are available from references 4–7. Reversed phase partition chromatography has been defined on p. 99.

TABLE 2.4

Solvents for Column Partition Chromatography

Stationary Phase	Mobile Phase
Normal Partition	
Water	Alcohols (n-butanol, isobutanol)
Water plus acid	Hydrocarbons (benzene, toluene, cyclo-
Water plus alkali	hexane, hexane)
Water plus buffer components	Chloroform
Aqueous alcohols (MeOH, EtOH)	Ethyl acetate
Alcohols (MeOH, EtOH)	Ethylene glycol monomethyl ether
Formamide	Methylethyl ketone
Glycols (Ethylene, propylene, glycerol)	Pyridine
Reversed Phase Partition	
n-Butanol	Water
Octanol	Water plus acid
Chloroform	Water plus alkali
Chlorosilanes and silicones	Water plus buffer components
Mineral oil	Aqueous alcohols (MeOH, EtOH)
Paraffin	Alcohols (MeOH, EtOH)
	Formamide
	Glycols (Ethylene, propylene, glycerol)

An experimental approach to this problem can be outlined as follows: prepare four thin layers of silica gel, cellulose, or kieselguhr (perhaps on microscope slides, p. 43) and place them in a desiccator over water overnight to impregnate them. Spot them with the sample to be investigated and develop them with n-butanol saturated with water, n-butanol saturated with $1N$ acetic acid, n-butanol saturated with $1N$ ammonium hydroxide,

[4] F. A. v. Metzsch, *Angew. Chem.* **65**, 586 (1953).
[5] F. A. v. Metzsch, *Angew. Chem.* **68**, 323 (1956).
[6] S. Berstrom and A. Norman, *Proc. Soc. Exp. Biol. N.Y.* **83**, 71 (1953).
[7] A. Norman, *Acta Chem. Scand.* **7**, 1413 (1953).

and ethyl acetate saturated with water. Visualize the chromatograms and compare the results. The developing solvents chosen were two neutral solvents, one acid, and one base. These systems can now be modified to produce the desired results. For example, suppose one wanted to modify the water-butanol system. Suppose further that the solutes move with the solvent front in the initial experiment. Hopefully, if the solutes can be retarded, they will separate. Their migration can be reduced by using acids or bases in the layers (p. 55) if they are acidic or basic in nature. Thus, bases would be retarded if the stationary phase had a pH of less than 7. The system can also be modified by adding a less polar solvent such as benzene to the moving liquid which could decrease its ability to dissolve the solute and, again, retard the migration. Systems developed in partition TLC are normally transposable to column systems.

A number of partition systems that have been worked out for the column-partition chromatography of various types of compounds are given in Table 2.5. These have been abstracted from a more extensive table in the book by Cassidy.[8] In the last column we have suggested some visualization techniques that would be suitable in TLC. The abbreviations are the same as those used in Table 2.3 on p. 31.

TABLE 2.5
Column-Partition Systems

Compounds	Stationary Phase	Moving Phase [a]	Visualization (TLC)[b]
On Silica Gel			
1. N-Acetylpeptides	H_2O	a. n-BuOH–$CHCl_3$ b. EtOAc–H_2O	H_2SO_4 and heat
2. C_2–C_{12} aliphatic acids	NaOH–MeOH (7.5 ml of $1N$ NaOH made up to 1 liter with MeOH)	a. isooctane–EtOEt (1:9) b. EtOEt	H_2SO_4 and heat
3. C_4–C_6 dicarboxylic acids	H_2O	n-BuOH–$CHCl_3$ (1:9) followed by (1:4)	bromcresol green
4. Aromatic di- and tribasic acids	H_2O	n-BuOH–$CHCl_3$ solutions (Increasing n-BuOH by gradient elution)	bromcresol green
5. Monohydric alcohols	H_2O	CCl_4 with increasing amounts of $CHCl_3$ and finally $CHCl_3$–HOAc (9:1)	H_2SO_4 and heat

[8] H. G. Cassidy, "Fundamentals of Chromatography," Interscience Publishers, Inc., New York, 1957, pp. 126–130. Original literature is cited therein.

TABLE 2.5 (cont.)

Compounds	Stationary Phase	Moving Phase [a]	Visualization (TLC) [b]
6. Aldehydes as semicarbazones	H_2O	$CHCl_3$–n-BuOH	H_2SO_4 and heat on silica gel or kieselguhr
7. 2,4-dinitrophenyl-amino acids	H_2O	n-BuOH–$CHCl_3$	colored compounds
8. Partially methylated glucoses	H_2O	n-BuOH–$CHCl_3$	H_2SO_4 and heat
9. Phenols and cresols	H_2O	cyclohexane	U.V. on phosphors
On Kieselguhr (Diatomaceous Earth)			
1. C_2–C_{10} aliphatic acids	27–35 N H_2SO_4	a. benzene b. benzene–pet. ether	heat
2. Di- and trihydric alcohols	H_2O	a. EtOAc b. benzene–n-BuOH	H_2SO_4 and heat
3. Penicillins	pH 5.5 citrate buffer	EtOEt–diisopropyl ether (1:1)	H_2SO_4 and heat
4. Pentose nucleosides and nucleic acids	H_2O	n-BuOH	U.V. on phosphors
On Cellulose			
1. Amino acids	H_2O	a. n-BuOH b. n-BuOH–HOAc–H_2O (3:1:1) c. phenol–H_2O (3:1)	ninhydrin
2. Sugars and derivatives	H_2O	n-BuOH	aniline phthalate or anisaldehyde
3. Methylated sugars	H_2O	ligroin–n-BuOH (3:2)	anisaldehyde

[a] It should be understood that the moving phase has been saturated with the stationary phase by shaking the two together in a separatory funnel.
[b] Directions for the preparation of these reagents are found on p. 69.

2.6. GAS CHROMATOGRAPHY

In gas chromatography the major points of variation are the nature of the stationary liquid phase and the temperature of the operation. These have relatively little in common with the concepts discussed in this chapter, and will be considered in detail in Chapter 5.

3

Thin-Layer Chromatography

3.1. INTRODUCTION

Thin-layer chromatography, or TLC, is the simplest of the techniques that will be presented in this book. The basic steps of the method have been previously described on p. 3 and illustrated in Figures 1.1–1.4. By use of this technique, separations of such widely differing substances as inorganic ions, inorganic-organic complexes, and synthetic or naturally occurring organic compounds can be achieved in a few minutes with equipment costing only a few dollars. Quantities as low as a fraction of a microgram or as high as five grams can be dealt with, depending upon the type of equipment used and the chromatographic phenomena involved.

Like most techniques, TLC can be used on several levels of complexity or sophistication. These levels are, in increasing complexity: microscope-slide TLC, macro-layer TLC, preparative TLC, and quantitative TLC.

TLC on microscope slides yields separations of mixtures of up to four components in about five minutes using normal laboratory equipment and a commercial adsorbent. The layers are easy to prepare, generally require no activation, and yield sharp separations. For these reasons, it is preferred by many investigators over the more complex techniques.

The next level of complexity is TLC on larger layers, layers measuring 5 × 20 cm, 10 × 20 cm, or 20 × 20 cm. These layers require some special equipment for preparation and longer development times (30 minutes to one hour). However, they can be used to separate mixtures containing a large number of components and are amenable to a number of special techniques of development such as two-dimensional TLC, TLC on shaped layers, continuous TLC, etc. The separated spots are somewhat easier to

39

visualize on the larger layers with special spray reagents. Quantitative TLC and preparative TLC are frequently carried out on layers of this size.

The third level of complexity, preparative TLC is a special technique for a special purpose. The layers are usually thicker and frequently larger than those used for other purposes. The major complications involve spotting techniques and methods for separating maximum quantities. TLC techniques are also most useful as an adjunct to column chromatography for preparative purposes.

The final level of complexity, quantitative TLC, is again a special technique for a special purpose. Although usually carried out on standard layers, the method involves several complications. These are: quantitative sample spotting, quantitative sample recovery, and the actual technique of quantification.

TLC involves essentially two variables: the nature of the stationary phase or layer, and the nature of the moving phase or developing solvent mixture. The stationary phase can be a finely divided powder functioning as an adsorbing surface (adsorption chromatography) or as a support for a liquid film (partition chromatography). Almost any powder can be and has been used as an adsorbent [1] in TLC, but we shall confine this book to the four most commonly used: silica gel (silicic acid), alumina (aluminum oxide), kieselguhr (diatomaceous earth), and cellulose. The moving phase can be almost any solvent or combination of solvents, and the choice of or search for a system that will resolve a given mixture is the most difficult problem in TLC.

Although we shall discuss four adsorbents in this text, it should be noted that probably 80% of the reported work on TLC has been done on only one, silica gel. Silica gel and alumina, when activated by heat, have a high adsorption capability and can be used to separate most of the less polar organic molecules. The developing solvents are organic and the chromatography is based on adsorption. Silica gel, kieselguhr, and cellulose, when properly coated with a liquid film, serve as supports for partition chromatography. As such, and with the appropriate solvent systems, they can be used for the separation of a large number of molecules, both polar and nonpolar. A number of additives such as binders, acids and bases, buffers, and visualizing agents can be incorporated into thin layers to serve specific purposes.

In the following sections the levels of complexity mentioned above will be discussed respectively. Section 3.2 will present a complete discussion of TLC on microscope slides and should suffice for work at that level. Section 3.3 will present a discussion of TLC on larger layers which, when taken with Section 3.2, will be complete. Sections 3.4 and 3.5 will involve the

---

[1] The stationary phase in TLC is frequently called the adsorbent, even when it is functioning as a support for a solvent in partition work.

additional complexities produced by working for specific types of results, preparative and quantitative; they will essentially be extensions of Section 3.3.

3.2. CHROMATOGRAPHY ON MICROSCOPE SLIDES

Most of the information in this chapter is taken from a remarkable paper by J. J. Peifer,[2] the Hormel Institute, Austin, Minnesota. TLC on microscope slides represents the simplest of the many "systems" that have been proposed. Although it can be used for quantitative work, it is probably best suited for crude qualitative work such as monitoring organic or biochemical reactions, exploratory work in searching for solvent systems, or in checking the efficiency of nonchromatographic separations. The major advantages are speed and convenience.

Adsorbents

Adsorbents for TLC are, in order of importance, silica gel, alumina, kieselguhr, and cellulose. They are much more finely divided (passing a 200-mesh screen) than are column adsorbents and generally contain a binder.

Silica Gel.   Silica gel is the most extensively used of all the TLC adsorbents and would probably be the best material to investigate first. If the compounds to be separated are neutral and contain one or two functional groups, they may be resolvable on layers of activated silica gel by using normal organic solvents or mixtures of solvents for developing. If the materials to be separated are organic bases, the developing solvent should contain a small amount of ammonium hydroxide or diethylamine. Conversely, a small amount of acetic acid should be added to the developing solvent when acids are to be separated (see p. 28). When the developing solvents contain water, the silica gel layers should not be activated (by heating) if cast from an aqueous slurry (p. 49). If prepared by dipping (p. 43), they should be hydrated by holding them over a beaker of boiling water. This will leave, or place, water in the layer to serve as a stationary aqueous phase.

Alumina.   Alumina is frequently used for the separation of bases rather than silica gel. Because of its basic nature, additional base in the developing solvent is not required. TLC on alumina is also useful as an adjunct to column chromatography (p. 85), where alumina is used much more extensively than silica gel.

Kieselguhr and Cellulose.   Both kieselguhr and cellulose are supports for liquid films in partition chromatography. This type of chromatography is always used for the separation of such very polar molecules as the amino

[2] J. J. Peifer, *Mikrochim. Acta,* 529 (1962).

acids, the carbohydrates, and various other naturally occurring hydrophilic compounds and is frequently required for the separation of closely related isomers (see Chapter 2). Care must be taken to assure that the layers contain a stationary liquid, generally water. This is especially true when the layers are to be prepared by dipping (p. 49). The solvent systems used with these adsorbents generally contain several constituents, one of which is the stationary liquid, for example, water.

Binders.   Thin layers are held in place on a glass plate by various binders. The most common of these binders is plaster of Paris (hydrated calcium sulfate), which is added to the adsorbents in amounts of 10–15%. Other common binders are starch (added to a level of 1–3%), low-molecular-weight silicon dioxide, and several organic polymers such as polyvinyl alcohol. Starch and the polymeric binders give very hard layers. However, the last two have a serious disadvantage; the developed chromatogram cannot be visualized by spraying it with sulfuric acid and charring (see p. 67). The instructions given in the next section for the preparation of microscope-slide layers use adsorbents containing plaster of Paris as a binder.

Visualizing Agents.   Two materials can be incorporated into microscope layers to assist with the visualization of the developed chromatogram: a phosphor and sulfuric acid. Of these the most useful is a phosphor.

Phosphors are substances which emit visual light when irradiated with light of another wavelength, generally one in the ultraviolet region. Thus, a layer containing such a substance will "shine" when viewed under U.V. light. If the compound to be visualized on the layer contains conjugated double bonds or is aromatic, it will quench this emitted light, probably by preventing the exciting U.V. light from reaching the phosphor. Such an effect results in the appearance of a dark spot or band on a bright layer. The method is quite sensitive and generally nondestructive. The most useful phosphors are inorganic compounds that do not contaminate the organic materials being chromatographed. They are usually added in amounts up to about 1% and generally emit light when irradiated with U.V. light with a wavelength of about 254 mμ. Many organic compounds fluoresce themselves when irradiated with U.V. light. This too is a useful property in TLC, but is much more rare and is not nearly so quantitative as the quenching effect described above.

The incorporation of sulfuric acid into microscope-slide layers is described in Table 3.1. After the chromatogram is developed, it is visualized by holding it over a hot plate with forceps. The residual sulfuric acid chars most organic substances, yielding black spots. The method is satisfactory for the chromatography of neutral or acidic substances, but generally does not work with bases since bases cannot be moved on the strongly acidic layers.

TABLE 3.1

Recipes for the Preparation of Peifer Slurries

Adsorbent	Slurry Medium	Proportions, g in ml
Silica gel G [a]	chloroform–methanol (2:1, v/v)	35 g in 100 ml
Silica gel G [a]-sulfuric acid	chloroform–methanol–sulfuric acid (70:30:2.5, v/v/v)	50 g in 102.5 ml
Cellulose powder [b]	chloroform–methanol (50:50, v/v)	50 g in 100 ml [b]
Alumina [c]	chloroform–methanol (70:30, v/v)	60 g in 100 ml [c]
"Florisil" [d]	chloroform–methanol–acetic acid (70:30:1, v/v/v)	55 g in 101 ml [d]

[a] Any of the commercial plaster-of-Paris bound adsorbents may be substituted. The adsorbents may, if desired, contain phosphors.

[b] Most cellulose powders will form satisfactory layers without plaster of Paris due to their fibrous natures. However, especially good layers can be prepared by triturating 35 g of cellulose and 15 g of plaster of Paris in a minimum of methanol and diluting the viscous paste to give the above ratio.

[c] 45 g of activated alumina and 15 g of plaster of Paris (or 60 g of commercial bound adsorbent containing this binder) are triturated with a minimum volume of chloroform-methanol and diluted to the above ratio.

[d] 45 g of "Florisil" and 10 g of plaster of Paris are triturated with 1 ml of acetic acid and a minimum volume of chloroform. The resulting paste is diluted to the above proportion. Although "Florisil" will not be discussed in detail in this text, it was an integral portion of Peifer's original paper.

Commercial Adsorbents.   As mentioned previously, adsorbents for TLC have a controlled particle size and contain binders and sometimes phosphors. Because of these complications, it is strongly recommended that commercial adsorbents for TLC be purchased and used rather than "homemade" mixtures. A number of these materials are described in Table 3.2 with their various additives and properties. The addresses of the suppliers are given in the Appendix.

Layer Preparation

Thin layers are most conveniently prepared on microscope slides by dipping the slides into a slurry of adsorbent in a mixture of organic solvents. Recipes for these slurries are given in Table 3.1. There will be some variations in the solid:solvent ratios depending upon the source and type of adsorbent. Within limits, thicker slurries will yield thicker layers. If the layers are thin and grainy, the slurries are too thin and more adsorbent should be added. The addition of sulfuric acid (to silica gel G) or acetic acid (to "Florisil") produces acidic layers. The slurries are stable for several weeks if kept in a tightly sealed container.

The following procedure should produce satisfactory layers.

(1) Prepare the slurry as described in Table 3.1. Shake it for about two minutes.

(2) Dip two microscope slides, held back-to-back, into the slurry as shown in Figure 3.1. Withdraw them slowly and allow them to drain on the edge of the container.

(3) Separate the slides and wipe the excess adsorbent off of the edges. Allow them to dry for about five minutes. Layers of silica gel and alumina prepared in this fashion do not require further activation. They should be prepared anew each day or stored in a dry atmosphere.

(4) If the layers are to be used for partition chromatography with water as a stationary phase (silica gel, cellulose, or "Florisil") they must be re-hydrated before use since the organic solvents in the slurry remove any water initially present. This can be accomplished by holding them over a beaker of boiling water and allowing them to dry at room temperature.

(5) After use, the microscope slides should be wiped off, washed with soap or detergent, and dried. For experiments dealing with very small quantities or in quantitative work where a slight contamination would be harmful, the slides should be rinsed with detergent followed by a 50% aqueous methanol solution acidified with hydrochloric acid.

In addition to the dipping technique, layers can be prepared on micro-scope slides by almost any of the more general methods discussed in Section 3.3.

Application of the Sample

The two major disadvantages of TLC on coated microscope slides are that the layers are relatively thin compared to macro-layers and that the distance available for chromatography is much less. Thus, it becomes im-perative to apply much smaller samples and to spot them in the smallest area possible. This can best be done with a fine capillary pulled from a glass tube such that it is not much thicker than a straight pin. The sample should be applied about 8–10 mm from the end of the slide, which is com-pletely coated (see Figure 3.2). Several applications can be made as long as the layer is dried between applications. As many as three samples can be placed on one layer if proper care is taken.

The mixture to be resolved can be dissolved in any convenient solvent for spotting. However, the best solvents for this purpose have boiling points between 50° and 100°. Such solvents can be handled conveniently, but will evaporate from the layer easily. The spotting solvent should be completely removed from the layer prior to chromatography, if necessary with a hot air gun or an electric hair dryer.

Choice of Developing Solvent

Solvent systems for TLC can be chosen by trial and error or by con-sultation with the literature. The simplest systems are mixtures of organic solvents used to separate mono- and difunctional molecules by adsorption

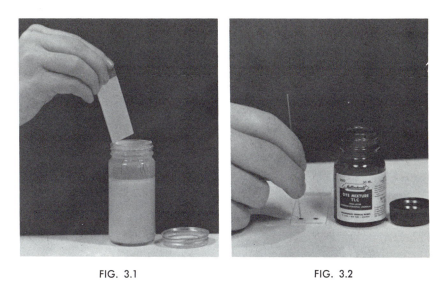

FIG. 3.1   FIG. 3.2

FIG. 3.3   FIG. 3.4

FIGS. 3.1 to 3.4. Two microscope slides, back-to-back, are dipped into a silica gel slurry (Fig. 3.1), separated, and allowed to dry. They are spotted (Fig. 3.2) with dye mixture (see Fig. 1.1) and developed with benzene (Fig. 3.3). The cellulose tape impression is taken from the dried, developed layer (Fig. 3.4) for recording in a laboratory book.

chromatography on layers of activated silica gel or alumina. The tendencies of a number of organic solvents to move materials over an activated surface have been measured, and the solvents have been arranged in the so-called eluotropic series shown in Table 2.1. The solvents at the bottom of the series are rather polar and move most compounds, whereas those at the top are nonpolar and will move few compounds. The polarities produced by blending solvents are shown in Figure 2.1. For most adsorption work, it is about as easy to search for a solvent system by trial and error guided by the eluotropic series as it is to consult the literature. A general discussion of solvent choice is given in Chapter 2, p. 28.

Although such empirical methods will generally solve a given problem, it may be well to consult the literature at some time during the investigation. The books listed in the Bibliography on p. 145 describe experimental conditions for a number of separations that have been obtained. Several of the suppliers of TLC equipment (Brinkmann, Gelman, Research Specialties, Pharmacia) provide up-to-date bibliographies on TLC and the *Journal of Chromatography* publishes a running bibliography on the subject in the back of each issue.

Development of Microscope-Slide Layers

The spotted chromatogram is placed in a small beaker or jar containing a layer of solvent a few millimeters deep as shown in Figure 3.3. The solvent level in the jar should be below the spots on the chromatogram. The jar is closed with a lid or a piece of aluminum foil and the solvent is allowed to ascend about three-fourths of the way up the layer. It is usually not necessary to saturate such small chambers (see p. 58).

Development requires about five minutes, depending upon the adsorbent and solvent. If the spots have an R_f of less than 0.5, the chromatogram should be redeveloped using the same solvent. This technique will always yield an improved separation. It is known as multiple development and is discussed in detail later (p. 60).

Visualization of Microscope-Slide Layers

Three methods of visualization are most commonly used on microscope-slide layers and are compatible with the simple approach given thus far. These are: iodine vapors, U.V. light on phosphor-containing layers, and sulfuric acid charring.

Visualization with iodine is carried out by placing the developed, dried layer in a jar containing crystalline iodine. After the jar is closed, the iodine vapor is adsorbed into the areas of the layer containing organic compounds, yielding brown spots on a white background. The spots become darker when left in the iodine, but generally fade rapidly when the layer is removed from the chamber. Most, but not all, organic compounds can be seen using this technique, and it is generally nondestructive as the iodine usually sublimes out of the layer leaving the compounds unchanged.

The visualization of chromatograms in U.V. light has been adequately discussed in the section devoted to adsorbent additives on p. 42.

One of the recipes for Peifer slurries (Table 3.1) involves the addition of sulfuric acid for the purpose of visualization. This has been discussed on p. 42. Alternately, one might spray a normal layer with concentrated sulfuric acid and place it in an oven at 110–150°. The technique reveals organic compounds as black spots and is destructive.

A large number of special spray reagents have been developed for detecting specific classes of compounds in TLC and paper chromatography. Some of these will be listed later in this text (p. 69). They are quite suitable for small as well as large layers.

Documentation

Microscope-slide chromatograms can best be recorded in a laboratory notebook by attaching the layer to transparent tape. The tape is pressed onto the visualized layer as shown in Figure 3.4. The tape and the portion of the layer which adheres to it are then removed and an additional layer of tape is placed on the back, covering the layer. The resulting "tape sandwich" is then placed in a laboratory notebook. Alternately, one can draw a picture of the chromatogram.

3.3. TLC ON LARGER LAYERS

Macro-layer TLC is generally carried out on three standard sizes of layers: 5 × 20 cm, 10 × 20 cm, 20 × 20 cm. Of these three sizes, the most frequently used are 5 × 20 cm and 20 × 20 cm. These larger layers have several advantages over microscope-slide layers. They are thicker, adhere to the glass plate a little better, and provide a larger area for chromatography. It is possible to investigate more samples simultaneously (4 on the 5 × 20 cm layers or 18 on the 20 × 20 cm layers) and the development track or zone is longer, permitting the resolution of more complex samples. Some special development techniques such as continuous development, two-dimensional chromatography, and chromatography in shaped areas are more applicable to the larger layers.

There are also some disadvantages. The larger layers are more difficult to prepare and make some type of commercial apparatus desirable if not mandatory. Since the layers are cast from an aqueous slurry, they must be heated for at least one hour to activate them for use in adsorption chromatography. Once activated, the layers must be stored in a dry atmosphere. This requires desiccators or special equipment. The layers also require longer development times.

Much commercial equipment has been developed to prepare and deal with large layers in TLC. Some of this will be discussed in the following sections, but no attempt will be made to mention everything that is available. The list of manufacturers and suppliers in the Appendix should, however, be reasonably complete.

The material presented in this section represents an extension of Section 3.2, and a knowledge of that section is assumed.

Adsorbents

As in Section 3.2, the discussion here will be limited to four adsorbents: silica gel, alumina, kieselguhr, and cellulose.

Homemade Adsorbents.   The following recipes have been given for the preparation of silica gel and alumina powders suitable for TLC.

Mangold [3] has stated that a reasonable approximation of a commercial silica gel for TLC can be prepared by mixing Mallinckrodt [4] silicic acid, 200-mesh, with 10–15% of its weight of newly calcinated calcium sulfate (plaster of Paris, prepared by heating reagent grade $CaSO_4 \cdot 2H_2O$ at 180° for 24–48 hours). A 200-mesh screen produces particles smaller than 0.12 mm. In similar fashion, alumina for TLC can be prepared by mixing Alcoa Activated Alumina,[5] 200-mesh, with about 5% of its weight of calcium sulfate.

Kieselguhr supports for TLC should contain about 15% of calcium sulfate, and cellulose powders generally form reasonable layers with no binder.

Commercial Adsorbents.   Commercial adsorbents for TLC are available in a confusing array. Table 3.2 lists some of them. It should be noted that adsorbents will vary from one manufacturer to another in such properties as flow rates, purities, and power of resolution. The properties of some of the binders and phosphors frequently added to layers have been discussed on p. 42.

One simple variation in the choice of adsorbents is the use of mixed adsorbents. Mixtures of silica gel and kieselguhr sometimes yield faster and better separations than either one alone.[6]

Cleaning Adsorbents.   Commercial adsorbents vary widely in their degree of purity. The impurities are generally iron (in silica gel) and organic compounds. For the chromatography of the inorganic ions it is necessary to remove iron from the adsorbents. The following procedure, freely translated from Seiler and Rothweiler,[7] can be used.

[3] H. K. Mangold, *J. Am. Oil Chemists' Soc.* **38**, 708 (1961).
[4] See Appendix for address.
[5] Aluminum Co. of America, see Appendix.
[6] R. D. Bennett and E. Heftmann, *J. Chromatog.* **9**, 353, 359 (1962).
[7] H. Seiler and W. Rothweiler, *Helv. Chim. Acta* **44**, 941 (1961).

"Five hundred grams of silica gel G is treated with 1000 ml of 6N hydrochloric acid (500 ml of concentrated hydrochloric acid and 500 ml of distilled water), stirred and allowed to stand. The supernatant liquid, colored yellow by the iron, is decanted and the silica gel is washed twice more with acid and then with three successive 1000 ml portions of distilled water. The adsorbent is then filtered and washed with distilled water until the filtrate is only slightly acidic. It is finally washed with 250 ml of ethanol and 250 ml of benzene and dried in an oven at 120°."

The material from the above treatment *now contains insufficient binder.* Two grams of plaster of Paris or one gram of starch are added to 28 g of purified silica gel to obtain a usable adsorbent.

When materials are to be recovered from the adsorbent, as in preparative or quantitative TLC, the adsorbents, commercial or not, should be prewashed with methanol. This can be accomplished by bringing a methanolic slurry to a boil and allowing it to stand a few hours at room temperature. The adsorbent is then removed by filtration and dried at 110° for an hour or so.

Preparation of Layers

Supporting Plates.   Large thin layers are usually prepared by casting a film of an adsorbent-water slurry onto a support plate and allowing it to dry. The supporting plates are generally pieces of plate glass with slightly beveled edges to prevent accidents. The plates can and have been prepared from Pyrex glass, stainless steel, and aluminum. Most of these are commercially available, but offer little advantage for normal work.

The glass plates should be washed with detergent, rinsed with distilled water, and dried before use. It is suggested that they be routinely wiped with a wad of cotton saturated with hexane just before the layers are cast.

Slurries.   The optimum thickness of a slurry used to cast thin layers will, within limits, depend upon the method of casting. If the slurry is too thin, it will run off the supporting plate, and if it is too thick, it will not flow out of the applicator. Most manufacturers suggest water:adsorbent ratios, some of which are given in Table 3.3. The adsorbent and water are generally shaken together in a stoppered flask, ground together in a mortar and pestle, or slurried in a blender (for cellulose layers). The slurry, especially if it contains plaster of Paris as a binder, becomes thicker with time, and a uniform mixing time of about 45 seconds is frequently suggested. For most purposes, the slurry should have the consistency of pea soup. One should not hesitate to deviate from the ratios suggested in Table 3.3 whenever it seems necessary, because there appears to be an appreciable variation in the amount of water needed, depending upon the previous history of the adsorbent.

TABLE 3.2

Commercial Thin-Layer Chromatography Adsorbents [a]

U.S. Source / Adsorbent	Brinkmann [b] (Merck, Machery-Nagel)	Gelman [c] (Camag)	Mallinckrodt [d]	Applied Science [e]	Alupharm (Woelm)	Bio-Rad	Reeve Angel (Whatman)
Silica gel, no binder	N, N/UV, N-HR, N-HR/UV	D-O, DF-O, DS-O, DSF-O	TLC-4, TLC-7, TLC-4F, TLC-7F	Adsorbil-2 AD-5	Silica gel Woelm	Bio-Sil A, three particle sizes	SG-41
Silica gel, calcium sulfate binder	G, GF, P, PF, G-HR, G-HR/UV	D-5, DF-5, DS-5, DSF-5	TLC-4G, TLC-7G, TLC-4GF, TLC-7GF	AD-1 AD-N-1	Silica gel G Woelm	Bio-Sil A, three particle sizes	
Silica gel, misc. binders	H, HF, H-HR, S, S/UV, S-HR, S-HR/UV			AD-3 AD-4			
Aluminum oxide, no binder		DS-O, DSF-O			Alumina Woelm, acid, basic, and neutral	Alumina AG4, acid; AG10, basic; and AG7, neutral	
Aluminum oxide, calcium sulfate binder	G, GF	DS-5, DSF-5			Aluminum oxide G Woelm	AG-4, acid; AG-7, neutral; AG-10, basic	

50

Adsorbent				
Aluminum oxide, misc. binders	H, HF			CC-41
Cellulose powder, no binder	300, 300-HR, 300-F	D-O, DF-O, DS-O, DSF-O		Cellex N-1 Cellex MX (microcrystalline)
Cellulose powder, calcium sulfate binder	300G, 300GF			
Kieselguhr, calcium sulfate binder	G			

[a] This table represents most of the major suppliers of adsorbents. Many other dealers handle adsorbents, but they generally purchased them from these suppliers. Adsorbents without binders are suitable for thin layers, generally because of a small particle size.

[b] Some of these can also be obtained from Kensington Scientific Corp. Brinkmann has a code for adsorbent types: "G" means calcium sulfate bound; "H" means a silicon dioxide binder; "S" means a starch binder; "N" is no binder; "F" and "UV" mean phosphor-containing; "HR" means extra pure.

[c] Camag has code for adsorbent types: "D" means TLC; "S" stands for a special quick-running adsorbent; "F" means a phosphor-containing adsorbent; "O" means no binder; "5" indicates a 5% calcium sulfate binder.

[d] The trade name for Mallinckrodt adsorbents is SilicAR. Mallinckrodt has a code for adsorbent types: "4" means a pH of 4 or an acidic adsorbent; "7" means a pH of 7 or a neutral adsorbent; "G" means a calcium sulfate binder; "F" means a phosphor-containing adsorbent.

[e] AD-1 and AD-N-1 contain a calcium sulfate binder. AD-N-1 contains silver nitrate as impregnating agent. AD-2 and AD-5 are pure silica gels having slightly different effective surface areas. AD-3 and AD-4 are mixtures of silica gel and magnesium silicate.

Layer Casting. Layers of almost any size and up to a thickness of 1 mm or so can be prepared with a glass plate, some masking tape, and a glass rod.[8] Layers of masking tape (up to five layers) are built up on opposite edges of a glass plate. The slurry, as described previously, is poured on the plate and leveled with a glass rod which is supported on the tape. The process is shown in Figure 3.5. The thickness of the tape

FIG. 3.5. Preparation of thin layers using tape and a glass rod. (Reproduced from Lees and DeMuria [8] through the courtesy of the authors and the Elsevier Publishing Co.)

layer determines the layer thickness. The tape must be removed before the layers are activated. The process, although simple, is not very efficient for the preparation of large numbers of layers.

A large amount of commercial apparatus is available for the application of thin layers to glass plates, and essentially all of it is satisfactory. Two of these applicators will be discussed: the Stahl-Desaga [9] apparatus and the

[8] This idea was first published by T. M. Lees and P. J. DeMuria, *J. Chromatog.* **8,** 108 (1962) and is the basis for the Mallinckrodt and Kontes systems, see Appendix.

[9] This apparatus is sold in the United States by Brinkmann Instruments.

Camag [10] apparatus. These were the first commercial applicators and they have been widely used. Furthermore, they represent the two general principles of application.

TABLE 3.3

Recommended Amounts of Water to be Added
to Various Commercial Adsorbents

Adsorbent	Supplier	Adsorbent:water W/v, g/ml	Casting Apparatus
Silica Gel G series	Brinkmann	30/60–65	Desaga
Silica Gel N series	Brinkmann	30/60–70	Desaga
Silica Gel H series	Brinkmann	30/80–90	Desaga
Silica Gel S series	Brinkmann	30/80–90	Desaga
Silica Gel P series	Brinkmann	30/75	Desaga
Silica Gel D and DS series	Gelman	D 50/100, DS 60/100	Camag
Silica Gel TLC series	Mallinckrodt	6/12–14	—
Silica Gel AD series	Applied Science	30/45	—
Silica Gel Woelm	Alupharm	30/45	—
Silica Gel SG 41	Reeve Angel	30/60	—
Silica Gel Bio-Sils	Bio-Rad	30/60	—
Aluminum Oxide G series	Brinkmann	30/40	Desaga
Aluminum Oxide H series	Brinkmann	30/80–90	Desaga
Aluminum Oxide P series	Brinkmann	30/30 [a]	Desaga
Aluminum Oxide Woelm	Alupharm	35/40	—
Aluminum Oxide D series	Gelman	20/65	Desaga
	Gelman	20/50	Camag
Aluminum Oxide DS series	Gelman	20/45	—
Aluminum Oxide AG series	Bio-Rad	30/30	—
Cellulose 300 & 300 G series	Brinkmann	15/90	Desaga
Cellulose P and MX	Bio-Rad	P 1/6	—
		MX 1/4.5	—
Cellulose CC 41 Whatman	Reeve Angel	30/60	—
Kieselguhr G	Brinkmann	30/60–65	Desaga

[a] This slurry should be shaken for 1 min, allowed to stand 15 min, and shaken again before use.

The Stahl-Desaga applicator is shown in use and in a close-up picture in Figure 3.6. A set of glass plates is arranged on a plastic holder and the applicator is passed over them, depositing a slurry film. The holder is designed so that it will hold plates of the three standard sizes (20 plates, 5×20 cm; 10 plates, 10×20 cm; or 5 plates, 20×20 cm) as well as mixtures of sizes. The applicator itself is available in many models, of which the most useful is the variable thickness model, S 11. The thickness of the

[10] This apparatus is sold in the United States by the Gelman Scientific Apparatus Co. and by A. H. Thomas.

layers can be varied up to 2 mm. The major advantage of the system is convenience and efficiency and the major disadvantage is that layers prepared in one application sometimes have different thicknesses. This is particularly serious with the small layers and is caused by using glass plates with different thicknesses placed beside one another or by the warping of the plastic holder. Apparatus of this general type is also available from Kensington Scientific Corp. and Applied Science Labs.

FIG. 3.6 The Stahl-Desaga apparatus in use and a more detailed picture of the applicator. The applicator shown in the less detailed picture is the old, nonadjustable model. (Reproduced through the courtesy of Brinkmann Instruments, Inc.)

Two additional companies, Quickfit and Shandon, manufacture equipment using the same principle, but they have designed plate holders that will level all the plates with respect to one another. Thicknesses are guaranteed to about ±0.01 mm. The Quickfit apparatus is shown in Figure 3.7.

FIG. 3.7. The Quickfit apparatus for preparation of highly uniform thin layers. The device is actually a plate leveling apparatus. (Reproduced through the courtesy of Quickfit, Inc.)

The other applicator, the Camag applicator, functions in quite a different way. It is shown in use in Figure 3.8. The glass plates are passed under a slurry reservoir to receive the film. The gate which determines the layer thickness rides directly on the plate moving through so that the thickness of the glass plate is unimportant. On the other hand, each plate must be pushed through individually and must be the correct width for the apparatus. Layers are quite uniform.

Modified Layers. A number of substances can be incorporated into thin layers or added to them to yield layers for specific purposes. In general, these are acids, bases, buffers, and complexing agents.

The pH of layers can be varied widely by using aqueous solutions of acids, bases, and buffers rather than pure water for the slurries. Such acids as oxalic acid (0.5N in slurry) or sulfuric acid (0.5N or see Table 3.1) can be used to produce layers that will separate acids and neutral substances while holding bases at the origin. Layers prepared from slurries containing sodium hydroxide (0.5N in slurry) can be used for the separation of bases. Such acidic and basic layers help to prevent the tailing of acidic and basic substances being chromatographed. The most frequently modified adsorbent is silica gel, and it should be noted that the rate of thickening of the slurry (p. 49) may be changed appreciably by such additives. Buffers of various types have been used for the separation of a number of acidic and basic substances.

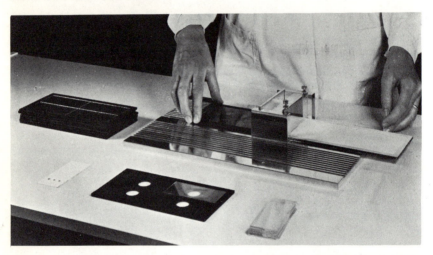

FIG. 3.8. The Camag apparatus in use. In this picture, a set of small slides has been placed in a holder and is being pushed under the reservoir. The system works equally well on large glass plates. (Reproduced through the courtesy of Camag A. G. and Gelman Scientific Co.)

The addition of complexing agents to thin layers for specific purposes is a most important concept. The three most commonly used agents are boric acid (0.1N in slurry) for the separation of sugars, silver nitrate (12.5% in slurry) [11] for the separation of olefinic materials, and 2,4,7-trinitrofluorenone [12] (0.3% added to adsorbent) for the separation of aromatic hydrocarbons. Each of these reagents complexes reversibly with the materials it is used to separate and enhances the resolving power of the layer markedly.

Layer Activation for Adsorption Chromatography. The cast slurry films

---

[11] Slurries containing silver nitrate should not be allowed to stand in stainless steel applicators. A plastic applicator is available from Applied Science Labs. and is strongly recommended.

[12] A. Berg and J. Lam, *J. Chromatog.* **16,** 157 (1964).

of silica gel or alumina should be allowed to stand for about 30 min and activated at 110° for a minimum of one hour. The adsorbent clinging to the edges of the glass plate should be wiped off and the layers should be stored in a dry-box or desiccator until used. The layers have a Brockmann activity of II–III (p. 89).

The layers can be sprayed with a saturated aqueous solution of silver nitrate to yield an impregnated layer. They should be reactivated after this treatment.

Layer Impregnation for Partition Chromatography.   Methods for layer impregnation will depend upon the liquid chosen for the stationary phase. When the liquid is to be water, the layers cast from aqueous slurries are simply allowed to dry at room temperature and heated for ten minutes at 105°. If the slurries contain added buffers, the layers are treated in the same manner and the result is a buffered stationary phase. Such systems account for the majority of partition chromatography.

Various other liquids can be used as stationary phases (p. 36). The impregnating liquids can be polar liquids such as formamide or glycols, or nonpolar liquids such as silicone fluid or paraffin oil. In general, the layers are activated to remove the excess water from the casting process and then impregnated with the desired liquid. The impregnating liquids can be added by dipping the activated layer (generally silica gel or kieselguhr) slowly into a solution of the liquid in a volatile solvent (20% formamide or ethylene glycol in acetone, 5% silicone oil [13] in ether, or 15% undecane in hexane). The layers are allowed to dry without heat.

Commercial Prepared Layers.   Prepared layers of almost any type are available commercially. They can be obtained on glass plates from Analtech, Brinkmann, Gelman, and Mallinckrodt, on plastic film from Eastman, and in the form of impregnated fiber glass from Gelman. All of the normal adsorbents are available, and the layers are of excellent quality. They are especially useful for quantitative work where an extremely uniform layer is required or where economic considerations make layer preparation prohibitively expensive.

Sample Spotting

The specific technique to be used for spotting samples on larger layers depends on the type of information or results desired. The amounts of substance to be applied will vary widely.

The use of capillaries is quite satisfactory for qualitative chromatography. If the concentration of the spotting solution is known (which is frequently not the case in TLC), it is desirable to know the amount of liquid applied. This can be done conveniently with the standard bore capillaries available from several manufacturers (Gelman, Helena Labs., A. H. Thomas,

---

[13] Dow-Corning, 200 fluid viscosity 10 cs.

Kensington, and others). These capillaries, when dipped in solution, will pick up known volumes ranging from 1 μl to 100 μl depending upon the bore size. When touched to a layer, they will deliver the liquid with a reasonable degree of accuracy. Otherwise, normal pulled capillaries with an outside diameter of about 1 mm are used. The spotting technique is shown in Figure 1.1 on p. 4. A microsyringe can also be used for spotting with a high degree of accuracy.

The amounts of sample applied range from 5 μg to 100 μg (5 μl to 100 μl of a 0.1% solution). It is generally desirable to investigate a sample at more than one concentration (on the layer). Thus, one might spot amounts of 5, 15, and 100 μg in three spots on a 5×20 cm layer. The lower quantities will generally yield sharper separations with less tailing and the larger quantities will reveal small amounts of impurities in the sample. The quantities spotted for partition chromatography are less (5–20 μg) than those for adsorption chromatography.

It is possible to spot certain types of compounds as their salts if a reagent is added to the developing solvent to regenerate them. Thus, base hydrochlorides can be spotted and will be chromatographed as free bases if a small amount (about 0.1%) of ammonia or diethylamine is added to the developing solvent.

The applications of samples for quantitative and preparative TLC are rather specific techniques and will be discussed later.

Solvent Choice

The discussion of solvent choice is given in Chapter 2, and very little needs to be added at this point.

One reagent has been added to developing solvents to act as a complexing agent. This is 2,5-hexanedione (acetonylacetone), which is added to various liquids in amounts of about 0.5% to aid in the separation of inorganic ions.

Development Techniques

Almost all TLC is carried out by the ascending technique shown in Figure 1.3. Many other techniques such as circular, horizontal, and descending development are known, but appear to show little special advantage. A number of topics and techniques, however, require some discussion in connection with the development process.

Chromatography Chambers and Chamber Saturation.   Thin-layer chromatography can be carried out in any convenient, sealable jar or container. For chromatography on larger layers, some attempt should be made to saturate the atmosphere in the chamber with solvent. This is usually done by lining the walls of the chamber with filter paper (halfway around and almost to the top). The paper should be wetted with solvent before the chromatography is started.

The so-called S-chambers or sandwich chambers offer several advantages. One commercial model is shown in Figure 3.9. In these chambers the glass plate holding the layer containing the sample to be separated is covered with a second plate. The two plates are held a few millimeters apart by a cardboard or Teflon spacer which is placed around three sides of the layer (the adsorbent around the edge of the layer is removed from the area under the spacer). The two layers are tightly clamped together producing an

FIG. 3.9. A sandwich chamber. (Reproduced through the courtesy of Brinkmann Instruments, Inc.)

intact chromatography chamber which is then dipped into a solvent for development. Commercial apparatus is available from Brinkmann, Gelman, Thomas, and Kontes. A special chamber of this type has been designed by Eastman for use with their prepared layers on plastic film.

Preequilibration of Layers.   In some cases, it is necessary or desirable to preequilibrate TLC chromatograms, that is, to allow the spotted chromatogram to stand in the presence of solvent vapor for an hour or so before it is developed. This is especially true for any type of partition chromatography since the stationary liquid should be saturated and in equilibrium with the moving liquid.

This operation can be accomplished experimentally by placing a second container, a dish or a small beaker, with a filter paper wick in the chromatography chamber. A portion of the developing solvent is placed in this container and the chromatogram is placed in the chromatography jar. After the period of equilibration, more of the developing solvent is poured into the bottom of the chamber and the chromatogram is allowed to develop. Solvent can be added through a hole in the chamber lid or by sliding the lid carefully to one side.

If adsorption chromatograms are allowed to preequilibrate, the resulting R_f values will be more repeatable.

Multiple Development.   The least appreciated and most important technique for the development of chromatograms is multiple development. In this technique, the chromatogram is developed, removed from the chamber, dried, and developed again in the same solvent. It is, in fact, a method for simulating an extended development, two developments to a height of 10 cm being equivalent to a single development of 17–18 cm. A major saving of time can be achieved, however, since the development rate decreases rapidly as the solvent moves up the layer and since the second development is generally more rapid than the first.

The technique can be treated mathematically and the following formula has been developed by Thoma [14] for predicting the optimum number of

$$n_{\mathrm{opt}} = \frac{-1}{\ln_e(1 - R_F)}$$

R_F = average R_f values of substances after one development.

developments for a given separation. In general, separations are always improved if the component spots are in the bottom half of the layer (R_f below 0.5) and are always damaged if the spots are in the upper third of the layer (R_f above 0.7).

It is possible to separate certain mixtures by several developments with a relatively nonpolar solvent system which *cannot be separated* with a single development using a more polar system. Much larger quantities can be separated using multiple development techniques, and consequently, these are quite important in preparative TLC (p. 75).

Chromatography in Shaped Areas.   It is sometimes desirable to shape the area used for TLC into the forms shown in Figure 3.10. This can be done with a spatula or sharp instrument. The spots are deformed as they come out of the narrow regions of the chromatogram so that they appear as narrow horizontal bands rather than as round spots. This sharpens separations and allows the separation of multicomponent systems. As shown in

[14] J. A. Thoma, *Anal. Chem.* **35**, 214 (1963); *J. Chromatog.* **12**, 441 (1963).

sample d of Figure 3.10, a small amount of impurity can easily be found if it happens to have an R_f lower than the major component.

Two-Dimensional TLC. Two-dimensional TLC can be used for the separation of mixtures containing many components having widely differing properties. A large area is available for the chromatography, and the two developments with two different solvent systems (in the two dimensions) allow the application of a wide range of conditions to the mixture.

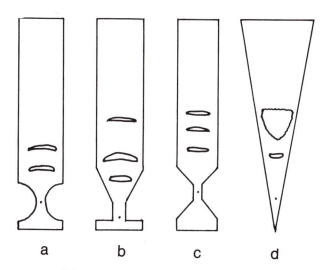

FIG. 3.10. Chromatography in shaped areas.

The sample is spotted in one corner of a square layer (20 × 20 cm) and developed with one solvent system so that the mixture is resolved in a track parallel to one edge (dotted circles in Figure 3.11). The layer is removed, dried, rotated 90°, and placed in a second solvent system so that the spots resolved during the first development are along the bottom and are again chromatographed. The resulting components (solid spots in Figure 3.11) can be located anywhere in a large area.

When the two solvents are different, the mixture is being submitted to two different resolving systems. The method is most useful for the separation of amino acids and was developed for their separation by paper chromatography.

The two solvent systems can be identical, but are normally different. The only reason for using identical solvents is to check a mixture for possible decomposition. If no decomposition has occurred, the spots will all be located on a line intersecting the original sample spot. The presence of any additional spots shows that some type of decomposition has occurred. This process is shown in Figure 3.12.

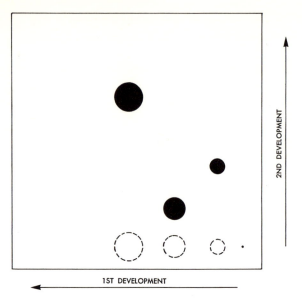

FIG. 3.11. An idealized two-dimensional chromatogram with different solvents used in the two dimensions. The dotted circles represent the position of the three components after the first development and the solid spots represent their location after the second development. There would appear to be only three components of the mixture.

Continuous Development. Continuous development over a long period of time offers one distinct advantage in that a relatively nonpolar solvent system can be used. A polar solvent system sometimes tends to override small differences between mixture components (p. 24). Continuous TLC can be carried out in the simple apparatus shown in Figure 3.13. In the apparatus the solvent evaporates as it comes out of the saturated chamber. The system will operate smoothly for several hours. Commercial apparatus is available from Brinkmann and Kensington (a descending, continuous system). The S-chambers described on p. 59 can be modified to serve as continuous development chambers. It is questionable whether continuous development offers much advantage over multiple development.

R_f *Values in TLC.* One of the major disadvantages of TLC is that R_f values are not very reproducible, particularly where adsorption chromatography is involved. Several factors seem to have some bearing on the situation.[15,16] These are (1) the adsorbent activity, (2) the chamber saturation, (3) layer preequilibration, (4) the conditions of development, (5) the temperature, and (6) the amount of sample.

[15] M. S. J. Dallas, *J. Chromatog.* **17**, 267 (1965).
[16] E. J. Shellard, *Lab. Practice* **13**, 290 (1964).

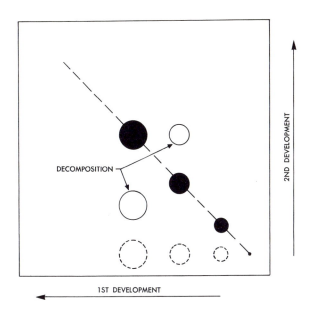

DECOMPOSITION

2ND DEVELOPMENT

1ST DEVELOPMENT

FIG. 3.12. An idealized two-dimensional chromatogram using the same solvent in both directions. The dotted circles represent the location of three components after the first development and the solid spots represent their final location. The solid circles represent decomposition products which have formed *in the course of the chromatograms.*

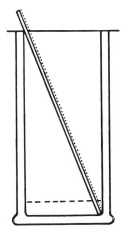

FIG. 3.13. A simple arrangement for continuous thin-layer chromatography.

63

The degree of adsorbent activity is reasonably standard when the layers have been activated at 110° for a minimum time of one hour. It is the time between activation and actual development which produces variations. If the layers are not stored in a dry atmosphere or if they are spotted on an especially humid day, the activity of the adsorbent is appreciably lowered and the R_f values are high.

Chamber saturation, layer preequilibration and development conditions can best be considered in the light of Figure 3.14. If a thin layer is placed in a chromatography chamber for development, several things happen as shown by the arrows. The solvent flows up through the layer; solvent vapor

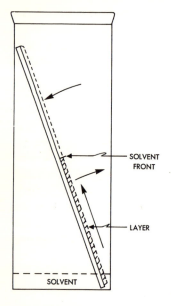

SOLVENT
FRONT

LAYER

SOLVENT

FIG. 3.14. A thin-layer chromatogram showing the various solvent movements. The solvent flows up through the layer, evaporates from it below the solvent front, and adsorbs into it above the solvent front.

adsorbs into the layer above the solvent front; and solvent evaporates from the layer below the solvent front. The R_f value observed will be dictated by the solvent flow through the layer as well as the state of the layer. The extent of evaporation from the layer will, however, alter the actual amount of solvent which must flow through the mixture to any given height and thus will affect the R_f. The adsorption of solvent into the layer will alter the adsorbent used during the latter stages of development. Proper chamber saturation and layer preequilibration as discussed previously, will minimize these effects and generally lead to lower, more repeatable R_f values.

Adverse effects will also be minimized if layers are developed a predetermined distance and are allowed to "overrun." Thus, a line of adsorbent should be scraped off at some standard distance from the origin so that each chromatogram is developed the same distance. The solvent is allowed to reach this line and the chromatogram is allowed to remain in the chamber ten minutes more ("overrunning"). The "overrunning" will allow the solvent to be more evenly distributed throughout the layer.

The temperature has a definite effect on chromatographic results. The effect is much more pronounced in partition chromatography than in adsorption chromatography since partition coefficients are quite temperature-dependent. Adsorption processes are less so. To obtain precise R_f values in partition chromatography, the temperature should be controlled to $\pm 2°$. For adsorption chromatography, the normal fluctuations of room temperature are permissible, although a drop of as much as $20°$ will appreciably lower R_f values.

The amount of sample has an effect on R_f values in adsorption chromatography. This depends upon the basic chromatographic properties of the substances being separated and is beyond control. Small samples give more valid R_f values than large samples since they more closely approach an ideal situation, where size is not a factor. R_f values may increase or decrease with concentration (see p. 20).

Frequently a substance in a mixture will behave somewhat differently from a pure sample, even on the same chromatogram. This may be due to a difference in sample size, but it is more likely to be due to the fact that an adsorbent can be slightly changed after a substance has moved over it. This change will be reflected in the R_f values of the substances that follow it. The presence of a given substance in a mixture should not be concluded from R_f values alone. The behavior on visualization should be considered. In addition, a small amount of the pure substance should be added to the mixture and chromatographed against the uncontaminated mixture and the pure substance. If the substance is indeed in the mixture, the appropriate spot in the contaminated or "spiked" mixture should be enlarged. This process is illustrated in Figure 3.15.

In summary, the following precautions should be taken when measuring R_f values for publication.

(1) Use standard commercial adsorbents.

(2) Always prepare and activate layers in the same way.

(3) Allow a minimum standard time between removal of a layer from the dry storage atmosphere and development.

(4) Apply small known amounts of the samples to be separated.

(5) Preequilibrate layers in a carefully saturated chamber or use an S-chamber.

(6) Develop chromatograms a constant distance and allow them to overrun.

(7) Report the average of three or more determinations.

The Problem of Tailing. Tailing, as the name implies, describes the situation when a sample spot is not round but has a tail something like a comet. Sample components overlap and clean separations are not achieved. The condition can generally be attributed to two factors. The first of these is the ionic character of acids and bases when chromatographed under neutral conditions. The ionic species of the equilibrium mixture (carboxylate ion for example) is much more polar than the nonionic species (the carboxylic acid) and will not move rapidly. As soon as these are separated, they reequilibrate, reseparate, etc. The inclusion of acids or bases in the developing solvents (p. 28) will do much to help this situation through a buffering action.

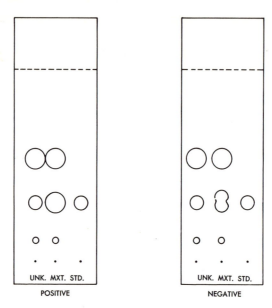

FIG. 3.15. Two chromatograms showing a positive identification and a negative result when an unknown is compared with a standard and a *mixture of the standard and the unknown.*

The second factor in tailing is the chromatographic character of the substances being separated. This situation is more apparent when large samples are involved and can be alleviated only by reducing the sample size. It is always a good idea to chromatograph a given mixture or sample at several concentrations (p. 58).

Visualization

Visualization techniques may be **destructive** or **nondestructive.** Destructive methods change the sample spots irreversibly and are useful in qualitative and some types of quantitative TLC. Nondestructive methods leave the sample components intact and must be used for preparative and some types of quantitative TLC. They can, of course, be used for qualitative work too. Visualization techniques may also be considered to be **universal** (valid for all or most compounds) or **specific** (valid for a given type or class of compounds).

One of the major advantages of TLC has been that such universal spray reagents as sulfuric acid or chromic acid can be used on the completely inorganic layers (plaster-of-Paris bound silica gel, alumina, and kieselguhr) to char any organic compounds to black carbon spots. This aspect has been somewhat overstressed. The use of selected, specific, visualizing agents can reveal much qualitative information about functional groups and general character of the substances being separated which is lost in the universal techniques.

Normally, thin layers are visualized by spraying them in a hood with an appropriate spray reagent from an atomizer bottle. Some reagents are commercially available in aerosol cans.

Universal Reagents or Techniques. Universal reagents are not really universal in that they work for all substances. Each technique has some limitations. The use of U.V. light on phosphor-containing layers, for example, is limited to those substances having conjugated double bonds (see p. 42). The iodine-chamber technique discussed on p. 46 works much better with unsaturated compounds than it does with saturated materials.

Several of the universal spray reagents commonly used in TLC are described in Table 3.4. They are all limited to organic compounds that are sufficiently nonvolatile to remain on the layer until they can be charred.

TABLE 3.4

Universal Spray Reagents Used in Thin-Layer Chromatography

Reagent	Composition and Use
Conc. H_2SO_4	Spray with acid and heat to $100-110°$ for a few minutes. Organic compounds will appear as black spots.
$H_2SO_4-Na_2Cr_2O_7$	Spray with a solution made by dissolving 3 g of $Na_2Cr_2O_7$ in 20 ml of water and diluting with 10 ml of conc. H_2SO_4. Heat at $100-110°$ for a few minutes. Organic compounds will appear as black spots.
$H_2SO_4-K_2Cr_2O_7$ (cleaning solution)	Spray with a saturated solution of $K_2Cr_2O_7$ in conc. H_2SO_4 and heat at $100-110°$ for a few minutes. Organic compounds will appear as black spots.
$H_2SO_4-HNO_3$	Spray with a solution of 5% HNO_3 in conc. H_2SO_4. Heat at $100-110°$ for a few minutes. Organic compounds will appear as black spots.
$HClO_4$	Spray with a 25% aqueous solution and heat to $150°$. Organic compounds will appear as black spots.
I_2	Spray with a 1% solution of I_2 in methanol or use I_2-chamber as described in text. Most organic compounds will appear as brown spots.

Of these universal techniques, the U.V. method is nondestructive. The iodine-staining technique is nondestructive for about 75% of organic compounds.

Specific Reagents. A few of the more important specific spray reagents are given in Table 3.5. They have been selected primarily because of their specificity as qualitative reagents and are destructive in nature. Many more can be found in the books listed in the Bibliography.

Documentation of TLC Results

One must have some method for recording chromatographic data in a laboratory notebook. The use of transparent tape to make an impression of a TLC chromatogram has been described on p. 47. The layer can also be sprayed with a plastic monomer which subsequently polymerizes and embeds the layer in a film. Commercial materials are available from Brinkmann (Neatan) and Krylon.

As an alternate procedure, a picture can be made of the chromatogram. This can be done using a Polaroid camera equipped with color film or any type of flat office copying apparatus such as a Xerox machine. As a last resort, pictures of the chromatograms can be drawn.

When R_f values are measured as recorded data or as data for publication, the R_f value of a common substance should be recorded under the exact conditions of the chromatogram. This will allow later comparison of chromatographic systems to the one reported.

3.4. PREPARATIVE METHODS

The separation of reasonable quantities of a mixture by any type of chromatography is difficult and frequently tedious. However, it is a reasonably sure method, if properly applied, whereas other methods are sometimes equally tedious and less certain to produce the desired results. All three of the chromatographic techniques discussed in this book have been used preparatively. The classic preparative method, column chromatography, is probably still the method of choice for the separation of quantities ranging from about 2 g to 100 g. Gas chromatography has been used for the preparative separation of both large and small samples, but the instrumentation required for large samples is quite complex. Preparative TLC is ideal for the separation of small samples (up to 2 g) of fairly nonvolatile compounds.

TLC concepts can be applied to preparative problems in two ways. The first of these is preparative TLC; the second is in the prediction of solvents for column chromatography and the analysis of column-effluent fractions. The former will be discussed here, and the latter will be discussed in Chapter 4 in connection with column chromatography.

TABLE 3.5

Specific Spray Reagents for Thin-Layer Chromatography

Reagent	Preparation and Use	Types of Compounds	Color
Aniline phthalate	soln. A: 0.93 g aniline and 1.66 g phthalic acid in 100 ml of *n*-BuOH saturated with H_2O a. Spray with A. b. Heat to 105° for 10 min.	reducing sugars	various colors
Anisaldehyde in H_2SO_4 and HOAc	soln. A: 0.5 ml of reagent in 0.5 ml conc. H_2SO_4, 9 ml of 95% EtOH and a few drops HOAc a. Spray with A. b. Heat to 105° for 25 min.	carbohydrates	various blues
Antimony trichloride in $CHCl_3$	soln. A: sat. soln. of reagent in alcohol free $CHCl_3$ a. Spray with A. b. Heat to 100° for 10 min. c. Observe in daylight and U.V.	steroids, steroid glycosides, aliphatic lipids, vitamin A, and others	various colors
Bromcresol purple	Spray with a 0.1% soln. of reagent in EtOH made just basic with NH_4OH or NaOH.	a. halogen ions, except F^{-1} b. dicarboxylic acids	yellow spots on purple
Bromcresol green	Spray with a 0.3% soln. of reagent in H_2O–MeOH (20:80) containing 8 drops of 30% NaOH per 100 ml.	carboxylic acids	yellow spots on green
2,4-Dinitrophenylhydrazine (2,4-DNPH)	Spray with a 0.5% soln. of reagent in $2N$ HCl.	aldehydes and ketones	yellow to red spots

TABLE 3.5 (cont.)

Reagent	Preparation and Use	Types of Compounds	Color
Dragendorff's reagent	soln. A: 1.7 g of bismuth subnitrate in 100 ml of H_2O–HOAc (80:20) soln. B: 40 g of KI in 100 ml H_2O a. Spray with soln. made from 5 ml of A, 5 ml of B, 20 g of HOAc, and 70 ml of H_2O.	alkaloids and organic bases in general	orange
Ferric chloride	Spray with a 1% aqueous soln. of reagent.	phenols	various colors
Fluorescein–Br_2	soln. A: 0.04% aqueous soln. of sodium fluorescein a. Spray with A. b. Observe in U.V. for conjugated systems. c. Expose to Br_2. d. Observe in U.V. for unsaturates.	unsaturated compounds	yellow spots on pink
8-Hydroxyquinoline–NH_3	soln. A: 0.5% soln. of reagent in 60% EtOH a. Expose to NH_3 b. Spray with A. c. Observe in U.V.	inorganic cations	various colors
Ninhydrin	soln. A: 95 ml of 0.2% reagent in BuOH plus 5 ml of 10% aqueous HOAc a. Spray with A. b. Heat to 120–150° for 10–15 min.	a. amino acids b. aminophosphatides c. amino sugars	blue
Silver nitrate–NH_4OH–fluorescein	soln. A: 1% soln. of $AgNO_3$ made basic with NH_4OH soln. B: 0.1% soln. fluorescein in EtOH a. Spray with A and B consecutively.	halogen ions	

In preparative TLC the sample to be separated is deposited in a thin line on one side of a large layer and developed in a direction perpendicular to this line so that the mixture will be resolved into bands. The bands are visualized nondestructively if they are not colored compounds, and the adsorbent containing the bands is scraped off the glass plate. The samples are then eluted from the adsorbent with a polar solvent. A typical chromatogram is shown in Figure 3.16, where a portion of each of the bands has been scraped off.

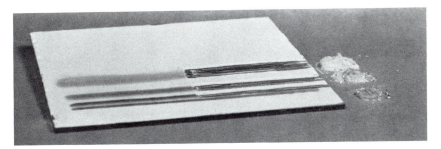

FIG. 3.16. A preparative thin-layer chromatogram of the dye mixture (see Fig. 1.1) on silica gel.

Quantities ranging up to 100 g have been separated by preparative TLC (using special equipment) but quantities less than 2 g are much more common. The technique is useful for separating reaction mixtures in order to obtain samples for preliminary study; for the preparation of pure analytical samples; in natural product structure work where small quantities are normal and mixtures are very complex; and for the preparation of pure samples for the calibration of quantitative methods (Section 3.5).

Each of the steps considered in the introduction to this chapter for TLC will be reexamined in respect to preparative work.

The Adsorbent

All of the normal commercial adsorbents can be, and have been, used for preparative work. As usual, silica gel has been used more than any of the others. A special series of adsorbents, the "P" series, has been marketed by Brinkmann for preparative TLC. Two factors are important in the preparation of adsorbents for preparative layers. The first of these is to make sure that the adsorbent is clean. If necessary, it can be prewashed with methanol as described on p. 49. The second is to use an adsorbent containing a phosphor (p. 42) whenever unsaturated or aromatic compounds might be encountered.

Layer Preparation

The optimum thickness for preparative layers is 1–1.5 mm. Thicker layers are hard to prepare and give poor separations. Such layers can be prepared with any of the commercial equipment discussed earlier in this chapter, although the Camag-Gelman apparatus is somewhat more efficient

FIG. 3.17. The Rodder apparatus for sample application to preparative thin-layer chromatograms. The tip of the fine needle of a near horizontal syringe is pulled over the layer while air pressure forces the sample out onto the layer. (Reproduced through the courtesy of Rodder Instruments.)

than the normal Desaga-Brinkmann spreader. Layers can be prepared using the edge-tape method discussed in Section 3.3 if the tape layers are made thick enough.

Sometimes, slightly thicker slurries are used to cast preparative layers, but this will depend upon the apparatus used. Generally, no problems are encountered in preparing layers up to 1.5 mm thick. The layers should be allowed to dry several hours at room temperature before activation. This will prevent cracking and casehardening. The activation is normal, that is,

110° for at least one hour. It is, in fact, advisable to store layers in a non-activated state and to activate them just before use.

Layers can be any size that is convenient to make and for which development chambers are available. The common size (20 × 20 cm) is most frequently used and will be the major one discussed in this text. Larger layers (20 × 100 cm and 20 × 40 cm) are used in several of the commercial systems (Brinkmann, Gelman, Shandon, Applied Science). Commercially prepared thick layers are available from Brinkmann and Analtech.

Sample Spotting

Sample spotting is the most critical step in preparative TLC. One must spot fairly large volumes of sample solution (up to 2 ml) in a thin (5–8 mm) uniform band without disturbing the layer. Any solvent boiling between 50 and 100° is suitable as a spotting solvent.

A number of pieces of apparatus have been developed commercially to carry out the spotting. The Rodder apparatus as shown in Figure 3.17 and the Applied Science apparatus as shown in Figure 3.18 are representative. Brinkmann supplies a device similar to the Rodder apparatus and Gelman-Camag manufactures one similar to the Applied Science model. Many additional devices have been described in the literature. In this text, we will describe a device that has been developed in our own laboratories.

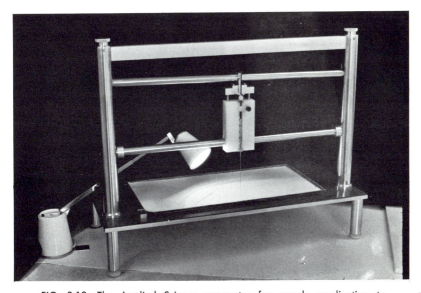

FIG. 3.18. The Applied Science apparatus for sample application to preparative thin-layer chromatograms. As the needle holder is moved from left to right, the slanting bar pushes the top of the syringe down, emptying the sample onto the layer. (Reproduced through the courtesy of Applied Science, Inc.)

A drawing of the streaking device is shown in Figure 3.19. In essence, it consists of a 1 or 2 ml pipette bent as shown with some type of wick or brush tip. The bends allow the rate of flow to be controlled by tipping the pipette. A wick tip can be made as shown in Figure 3.19 by looping a piece of fine nichrome wire and tucking the ends into the tip of the pipette. A piece of string or yarn is then tied into the loop and the loop is pushed into the pipette so that the string appears to emerge from the tip. The string is cut to a length of 7 or 8 mm. Alternatively, a brush tip can be fashioned from a small soft (no. 3) camel's hair brush. The wooden stick is worked out of a normal brush and the brush is soaked in methanol for

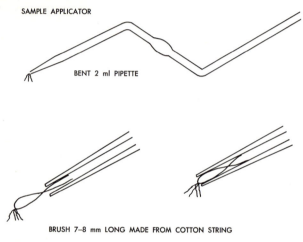

SAMPLE APPLICATOR

BENT 2 ml PIPETTE

BRUSH 7–8 mm LONG MADE FROM COTTON STRING

FIG. 3.19. Detailed drawing of a 2 ml pipette modified to serve as a device for sample application to preparative thin-layer chromatograms.

a few days to remove the rosin, which embeds the bristles, and to render the brush porous. The metal brush holder is then placed over the tip of the pipette and, if necessary, fastened with epoxy glue. The applicator is filled by sucking the solution up into the pipette in the normal way, and the sample is applied as shown in Figure 3.20. The position of the pipette in Figure 3.20 is quite important since it allows the rate of flow to be controlled by tipping.

Several successive applications will normally have to be made, and the layer should be dried with a hot air gun between applications. The use of freshly activated layers as suggested previously (p. 73) which are still warm, *but not hot,* will help to evaporate the spotting solvent.

In an alternate manner, the sample solution can be applied fairly uniformly to one edge of the layer with a syringe or pipette without attempting to keep it in an especially thin band. The chromatogram is then developed a short distance with a very polar solvent such as methanol so that the sample is compressed into a thin line. The layer is then removed, dried, and developed in the normal fashion.

FIG. 3.20. The device from Fig. 3.19 shown in use. It is necessary to hold the device near horizontal as shown so that the sample will not come out too fast.

Sample Size

The amount of a mixture that can be separated on a layer of a given size and thickness will vary widely, depending upon the type of compounds being separated and the ease of the separation. Reasonably safe quantities for separation on 20×20 cm layers 1 mm thick are 50 mg when adsorption chromatography is involved and 5 mg when the chromatography is partition. Quantities ranging up to 250 mg and 50 mg respectively have, however, been successfully separated under ideal conditions. For the separation of large quantities, a number of small layers or some type of oversized layers can be used.

Development

The development of preparative layers is normal. The same solvents can be used on both thick and thin layers and the separations are similar. The chamber should be well saturated. For best results, solvents should be chosen which will not move the mixture components above an R_f value of 0.5 in a single development, but which will move most of them out of the origin. This will allow a multiple development and will assure that even the slowest components have been separated from any residue that might remain at the origin. Larger samples can be separated by several developments with a less polar solvent than with a single pass of a more polar solvent. The layer should be dried with a hot air gun or in a warm forced-draft oven between developments.

Visualization

The visualization of preparative chromatograms must be nondestructive. Two of these techniques, U.V. light on phosphor-containing layers and the iodine staining technique, have been considered previously (on p. 42 and p. 46). Three additional methods are pertinent: water-spray, edge-spray, and transparent tape visualization. When a normal silica gel layer is saturated by spraying it with water, it becomes semitranslucent (like freshly cast slurries). Certain types of compounds, mainly saturated structures such as steroids and lipids, appear on these layers as opaque spots or bands. These can be marked while the layer is wet, and the separated components can be recovered after the layers have been dried.

The edge-spray technique is less desirable in that one must sacrifice a portion of the sample. A clean glass plate is placed over about 80% of a developed layer in such a way that portions of the origin line and the separated bands are left uncovered. A small amount of adsorbent is scraped off along the edge of the cover glass so that spray reagent cannot seep under it. The edge is then sprayed with any convenient spray reagent and the results are extrapolated across the layer. The protected portions of the bands of adsorbent containing the desired components are then removed and eluted.

Finally, a portion of the layer can be removed by pressing a piece of transparent tape across the bands (Figure 3.4 on p. 45). The tape can then be sprayed with an appropriate reagent or exposed to iodine vapor and used as a key to locate the bands on the chromatogram.

Sample Recovery

The adsorbent bands containing the (hopefully) pure components are then scraped off of the glass plates with a spatula, a razor blade, or a rubber policeman, generally onto waxed paper or aluminum foil. The adsorbent is placed in a sintered glass funnel (medium or fine grade) or in a filter paper cone in a glass funnel and extracted several times with an appropriate solvent. This solvent should be just polar enough to remove the sample without extracting any extra impurities from the adsorbent. Generally, a solvent is chosen which will move the sample to an R_f value of 0.8 or 0.9. If any doubt exists about the efficiency of a given elution solvent, the adsorbent should be extracted with methanol or methanol–ammonia, 9:1, before it is discarded. The solvent is evaporated from the combined eluents (perhaps keeping the final methanol washings separate), and the products are isolated. Generally, a final recrystallization is needed.

A Word of Caution

Almost all organic compounds decompose when allowed to remain on adsorbent layers for extended periods of time. They are present in an amorphous state, completely exposed to light and air. Preparative TLC should be carried out as quickly as possible from the initial sample spotting to the final elution. Each mixture should be checked for decomposition using the two-dimensional technique described on p. 63 (Figure 3.12) before preparative TLC is commenced, in order to see just how serious the situation might be. TLC can, when necessary, be carried out in inert atmospheres (nitrogen or argon) in the absence of light.

3.5. QUANTITATIVE TLC

Quantitative TLC requires technique and precision well above that discussed in the preceding sections of this chapter. Both instrumental and noninstrumental methods are available. As a quantitative technique, TLC is definitely inferior to gas chromatography when the compounds being assayed are volatile or can be converted quantitatively to volatile derivatives. Column chromatography is a poor quantitative technique. However, for nonvolatile substances or where the necessary instrumentation is lacking, TLC can be quite useful.

Two basic techniques are available for quantitative TLC. In the first technique, the substances to be determined are assayed directly on the layer. In the second technique the substances are removed from the layer and assayed, generally spectrophotometrically.

All quantitative work requires pure adsorbents and solvents. It may be well to prewash adsorbents before use (p. 49). Alternately, the layers may be prewashed by developing them with a polar solvent and drying them before use.

Assay on the Layer

When substances are assayed directly on the layer, there are no extraction or transfer errors and the procedures are quite simple. Unfortunately, these procedures are not extremely accurate and the overall error varies between 5 and 10%. Two methods of this type will be discussed: spot-area measurement and densitometry. Densitometry requires an instrument (a densitometer) but spot-area measurement does not.

Spot-area Measurement.   Quantification by spot area has been carried out by a number of workers, but the most comprehensive work has been done by Purdy and Truter.[17] These authors found that the square root of

[17] S. J. Purdy and E. V. Truter, *Analyst* **87,** 802 (1962).

the area of a spot is directly proportional to the logarithm (to base 10) of the weight of substance present. The proportionality constant is different for different substances and the relationship appears to be valid on quantities ranging from 1 to 80 μg on silica gel layers. A pure sample of the substance being determined must be available for calibration of the method (this can be obtained by preparative TLC).

The direct relationship cited above makes it possible to assay a substance without preparing a calibration curve. However, the preparation of such a curve gives some idea of the general accuracy and limits of the method. The following procedure is suggested.

(1) A solution containing a known concentration of the pure substance to be assayed is prepared and diluted twice so that three solutions of known concentration are available. These concentrations should be in the general range of the concentration of the substance in the sample to be analyzed; at best between 0.1 and 1%. Most solvents can be used to prepare these solutions. The exception is chloroform, which gives poor results.[17]

(2) The four solutions (the unknown and the three known solutions) are spotted in duplicate on a single large layer (20 cm × 20 cm). The spotting technique is most important and delicate. Since the spot size will be the crux of the method, it is necessary to spot the *same volume* of each solution in such a way that the spots are initially the same size and the layer is not disturbed (leading to spot distortions). This can best be accomplished with a microsyringe mounted in a ring stand so that the tip of the needle is just over the layer. The required amount of solution (5–10 μl) is squeezed out of the syringe and the *layer is raised* to accept the drop. Multiple applications should not be made.

(3) The chromatogram should then be developed a premeasured distance (10–12 cm) in a well saturated chamber. The layers are dried and visualized by any appropriate technique, including charring with sulfuric acid (10 min at 160°). An idealized chromatogram is shown in Figure 3.21.

(4) A piece of transparent paper is then placed over the layer and the spots are traced. The tracing is placed over a piece of millimeter graph paper and the squares in each spot are counted. A drawing instrument known as a planimeter (p. 136) can be used for this area measurement, but the square-counting process is less tedious than it sounds. Alternately, a photocopy such as a Xerox copy can be made of the visualized chromatogram, and the spots can be cut out and weighed to determine their area.

(5) The data from the three known solutions are then plotted on graph paper as the square root of the spot area vs. the logarithm of the sample weight. A straight line is drawn through the three points. The actual position of the points with respect to the line gives an idea of the accuracy of the method and the technique. The unknown sample is then determined using this calibration line.

(6) The slope of the line will remain reasonably constant for subsequent determinations, but the intercepts may change slightly from time to time, depending upon the exact conditions of the chromatography. If, with each additional analysis, a known sample is chromatographed, it can be used to determine whether a correction factor is needed.

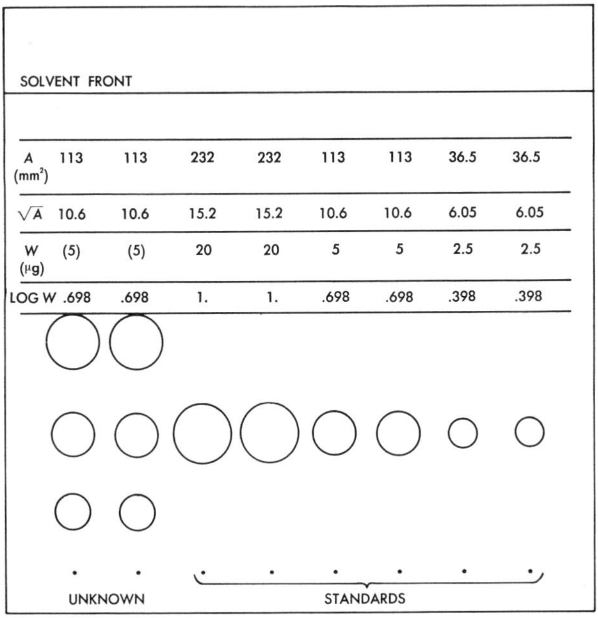

A (mm^2)	113	113	232	232	113	113	36.5	36.5
$\sqrt{A}$	10.6	10.6	15.2	15.2	10.6	10.6	6.05	6.05
W (μg)	(5)	(5)	20	20	5	5	2.5	2.5
LOG W	.698	.698	1.	1.	.698	.698	.398	.398

FIG. 3.21. An idealized chromatogram showing a quantitative determination by spot area. The center spot of the unknown is being determined. All of the samples were 5 μl. The standard solutions were 0.2%, 0.1%, and 0.05% and the "unknown" is 0.1% also. Duplicate samples are shown, and the actual data are tabulated and superimposed on the top of the picture with units which are convenient for a determination.

The algebraic method of Purdy and Truter [17] does not require a calibration curve. In the method, a standard solution of the substance to be assayed is again required. The solution containing the unknown is diluted, precisely, so that two related unknown solutions are available. The standard solution and the solutions containing the unknown are then chromatographed on the same layer, and the spot areas are determined as described above. The results are then calculated using Equation (1) where W and A are the weight and spot area of the unknown, W_s and A_s are the weight and area of the standard solution, A_d is the area of the diluted unknown spot, and d is the dilution factor (this factor is $\frac{1}{2}$ if the unknown solution is diluted with an equal amount of solvent). Convenient units are micrograms and square millimeters. One μl of a 0.1% solution contains 1 μg.

$$\log W = \log W_s + \left(\frac{\sqrt{A} - \sqrt{A_s}}{\sqrt{A_d} - \sqrt{A}}\right) \log d \qquad (1)$$

Densitometry. Quantification of a TLC chromatogram by densitometry is carried out by passing a developed, visualized chromatogram through the beam of a densitometer in such a way that the relative size and density of the spots can be measured. The spots on such a chromatogram are usually visualized with sulfuric acid charring, and the data from the densitometer are usually plotted on a strip-chart recorder. The areas under the curves of the chart paper are proportional to the amounts of substance present in the spots being measured. The technique is quite similar to the quantification of gas chromatographic results as discussed more completely in Section 5.5. Commercial models of densitometers are available from Photovolt, Inc. An idealized chromatogram and the strip-chart which might result from it

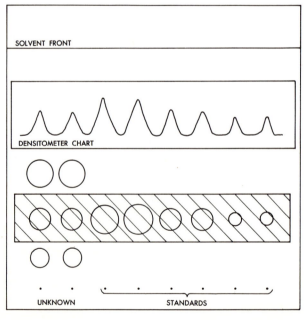

FIG. 3.22. An idealized chromatogram showing a quantitative determination by densitometry. The solutions and the quantities of solutions are the same as those in Fig. 3.21. The area covered by the diagonal lines was scanned by the densitometer. The areas under the peaks on the chart are directly proportional to the quantities of material present.

are shown in Figure 3.22. The area under the diagonal lines was scanned. The system illustrated is the same one that was used in the spot-area determination (Figure 3.21).

There are several complications inherent in the method. Slightly volatile compounds may partially sublime before they can be charred. Different compounds have different relative percentages of carbon, thus requiring a calibration curve for each compound being assayed. Very uniform layers are required so that the base value on the densitometer (the background) will remain constant. Commercially prepared layers are ideal for this purpose. The errors in the method can be as low as 5%, but will differ widely depending upon the type of compounds and the proficiency of the worker.

Assay by Elution Methods

In this second technique, one must contend with errors of transfer and extraction, but the ultimate procedures of assay are extremely accurate. Comparatively, the two techniques (assay on the layer and assay by elution) are about equal in overall accuracy. The assay is carried out by locating the spot to be assayed (by a nondestructive visualization method), removing the adsorbent containing the spot from the glass plate, eluting the sample from the adsorbent, and measuring the amount of material present by ultraviolet spectrophotometry or some specific colorimetric method. The following procedure is suggested.

(1) The mixture containing the substance to be determined is chromatographed on a large layer (20 × 20 cm) along with three solutions of pure substance of known concentration, in much the same manner as for the spot-area method. The pure samples will establish a calibration curve.

(2) The amounts of sample spotted should be known as accurately as possible and it is suggested that a microsyringe or microburette be used. However, the spotting technique itself is not especially critical.

(3) The chromatogram is developed in a normal fashion and visualized nondestructively (see p. 76). The best of these visualization techniques involves U.V. light on phosphor-containing layers.

(4) The chromatogram is marked with a sharp instrument into rectangular areas, each of which contains a spot of either the unknown or the standard. An idealized chromatogram is shown in Figure 3.23. One empty area is marked off to serve as a blank value. The rectangles should have areas as nearly identical as possible.

(5) The adsorbent within each rectangle is removed from the layer as completely as possible with a razor blade or the small vacuum-cleaning device shown in Figure 3.24.

(6) The sample is then eluted completely from the adsorbent with a polar solvent that will not interfere with the assay technique. The vacuum-cleaning device in Figure 3.24 is designed so that it can be used for the

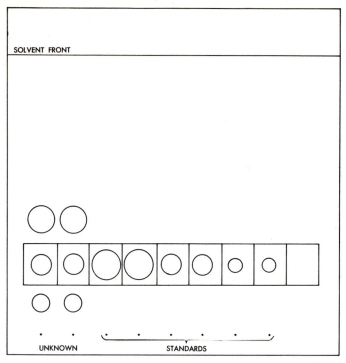

FIG. 3.23. An idealized chromatogram marked off for quantitative de-termination by elution methods. The area at the extreme right is the blank. Note that the areas of the rectangles are as nearly the same as possible. The solutions and the quantities of solutions used are the same as those in Fig. 3.21.

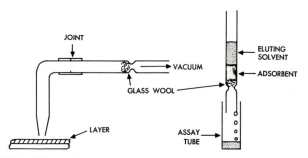

FIG. 3.24. A device for removing adsorbent from a layer for quantitative assay by elution. After the adsorbent has been picked up from the layer as shown on the left, the device is dismantled at the joint and used as an elution tube as shown on the right.

82

elution step as shown. Sometimes, one can simply stir the adsorbent with solvent, centrifuge it, and analyze the supernatant liquid.

(7) The eluents are then made up to a known volume and analyzed by any appropriate technique. This procedure provides calibration and blank values that will not need to be done for each analysis unless the adsorbent or the chromatography is altered.

Other Methods

Radioactive samples have been measured by counting techniques, both on the layer and after elution. Fluorescent substances have been measured on layers.

4

Column Chromatography

4.1. INTRODUCTION

Column processes are the more useful chromatographic methods for the separation of pure compounds, in quantity, from mixtures. Separation of the various solutes from a mixture occurs in an appropriate system as a result of differential migration of bands through a column of stationary phase. Very crude mixtures will ordinarily yield fractions composed of several similar solutes, whereas simple mixtures can frequently be manipulated to yield fractions composed of single components. In all applications of column chromatography, the mixture to be separated is introduced as a narrow zone at the top of the column of adsorbent or support material (see Chapter 1, Figures 1.6–1.9). Solvent addition causes a downward migration of the mixture of components, and the bands are collected as fractions as they emerge from the bottom of the column. A graphic presentation of the series of events is given in Figure 4.1.

The rate of travel for each component is determined by the combination of forces that control the adsorption or partition characteristics of the system. Column **adsorption** separations depend upon interaction among the adsorbent surface, the solute(s), and the solvent. Column **partition** chromatography is a process involving solute distribution between a mobile phase and a stationary liquid phase held as a film in a column of some hydrophilic solid support. Resolution of mixtures depends, therefore, upon successful manipulation of the forces which control these interactions. The nature of these forces and their manipulation in various situations have been discussed in detail in Chapter 2.

The first step in the design of a column chromatographic process must be the choice of the system—whether adsorption or partition (Chapter 2).

84

From a practical viewpoint, an investigator is normally dealing with a mixture whose composition has not been described, and the mixture is, in reality, an unknown. However, no reaction mixture or tissue extract is an absolute unknown, and the beginning point for the elaboration of a chromatographic process must be the known or *predictable* components of the mixture. In addition, with certain preparations, knowledge gained in prechromatographic procedures will provide some insight into the nature of the solute mixtures.

Since most mixtures contain some principles in larger quantities than others, it may be advantageous, as a first step, to subject such products to some prechromatographic fractionation to separate their major components. Such prechromatographic fractionations can also be used to separate mixtures into fractions which contain similar substances (for example bases, weak acids, strong acids, neutral components, etc.). There is little advantage in separating nonpolar components from polar components or acids from bases by a chromatographic fractionation when it can be readily accomplished in a separatory funnel using the appropriate extraction solutions and solvents. Such removal of the major components or prefractionation will allow the chromatography to be used most efficiently.

As a second step, it is recommended that the investigator consult the literature for some reference which might bear directly upon his problem. The reports of chromatographic separations are numerous (Chapter 6), and it is most probable that a useful system can be extracted directly from the literature or developed as a result of some modification of a published method.

TLC Relationship

A third step in the elaboration of a column system (or an initial step as an alternate to the first two steps), is to use data derived from preliminary experiments conducted in TLC systems (see Chapter 2). Such experimentation will reveal or confirm the comparative effectiveness of adsorption vs. partition systems and will frequently suggest an ideal solvent or solvent system for the development of the chromatogram. TLC data will also provide preliminary details concerning the complexity of the solute mixture. It is somewhat logical to assume that TLC data might be readily extended to column systems using equivalent materials. Such is ordinarily the case when *exact* equivalents in adsorbents are used. In that era when paper partition chromatography was widely used as a detecting process, the solvent systems used successfully on paper strips were readily extended to column systems of powdered cellulose. These preparative transpositions were conducted without serious deviation from the anticipated result. However, for adsorption chromatography no two brands of alumina or silica gel are exactly equivalent even though they may be quite identical in all measurable respects (activity grade, particle size, pH, etc.). Thus an exact equivalent means that the TLC adsorbent and the column adsorbent must come from the same commercial source. Several manufacturers who supply both TLC and column materials are noted in Table 4.1. The only variation that can be tolerated is particle size, which seems to be relatively unimportant. The normal particle size for TLC adsorbents (finer than 200-mesh) is too fine for successful column work. The solvent flow is impeded. Adsorbent sizes for column work are discussed on p. 89.

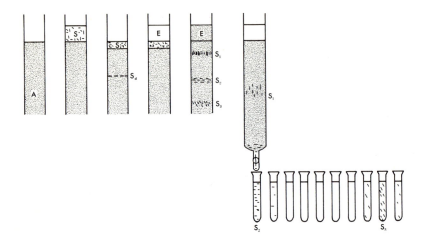

FIG. 4.1. Idealized column chromatography of a mixture of solutes, S_1, S_2, and S_3. The column is filled with adsorbent, A, and the sample, S, is added in solution. S_I is the distance reached by the sample solvent when the sample has been completely applied, and E is the eluting solvent. As the solvent leaves the column as the effluent, it is divided into fractions in the tubes shown.

In developing a TLC system as a model for a column process for preparative purposes it is necessary that the majority of solutes show an R_f value below 0.30. If there are a significant number of substances which have values above this it is most probable that the column system will not effectively resolve the mixture. This requirement is due to the reciprocal relationship that exists between the R_f value on a layer and the volume of solvent that will move a band of solute out of a column. High, similar R_f values (0.8 and 0.9) will not produce a sufficient difference in the solvent volumes to yield a separation, whereas low, similar R_f values (0.1 and 0.2) will.

The following procedure will allow the transposition of TLC conditions to column systems in the majority of cases involving adsorption chromatography.

(1) Find a solvent system that will yield a separation on a thin layer.

Ideally, such a system should consist of two liquids, one more polar than the other.

(2) Reduce the polarity of the solvent system until the components of the sample mixture have R_f values below 0.3 on the thin layer. It makes little difference whether a clean separation can be *seen* at this stage or not since it is known from step (1) that a separation can be achieved.

(3) Use this modified solvent system as a slurry liquid to pack a column of adsorbent as nearly identical to the TLC adsorbent as possible. (See p. 94 for further details.)

(4) Apply the sample to the column and develop it. (See p. 94 for further details.)

(5) Divide the column effluent into fractions and analyze them by TLC. The fractions may or may not overlap, depending upon whether the column was overloaded. (See p. 98 for further details.)

(6) In many cases in synthetic organic chemistry, only one spot will have a high R_f value (above 0.3) in the initial TLC experiments. In this case, the solvent system can be used without modification with a reasonable chance of success. Such a system is ideal for the separation of a desired product from tars and residues which will remain at the origin on a thin layer.

When partition chromatography is involved, the solvent systems that will produce a separation on TLC can frequently be transferred directly to a column. Here also, however, the separations should be made in the bottom half of the thin layer.

4.2. COLUMN ADSORPTION CHROMATOGRAPHY

The essential understandings regarding the nature of the adsorption process and the phenomena associated with solute separation in adsorption systems has been presented (Chapter 2). Additionally, those factors that determine which chromatographic technique is to be used have also been discussed. This section will be concerned with the materials and methods of column adsorption chromatography.

Columns

Chromatographic tubes are usually made of glass and are of various types (see Figure 4.2). They are intended to support the adsorbent and are designed to permit control of solvent input and effluent collection. The dimensions are variable but usually the length is at least 10 times the internal diameter and it may extend to 100 times that value. The width/length ratio is largely determined by the ease or difficulty of the separation. The size of the column and the amount of adsorbent used are largely determined by the weight of solute material to be resolved (see p. 89).

It is not always practical to completely elute all of the solute bands from a column. In fact, in one type of column chromatography (p. 99) the column is developed so that the solutes are distributed in bands over the length of the adsorbent. In these cases the column must be of such a design that the adsorbent, with the bound solutes, can be extruded or pushed out more or less intact. The columns, type (a), with a standard taper joint, or type (b), with an o-ring seal, are useful for this purpose. After the chromatogram has been developed, the joint connection is separated and the column of adsorbent is extruded for examination, segmentation, or analysis. These two types of columns are fitted with a fritted glass disc, and

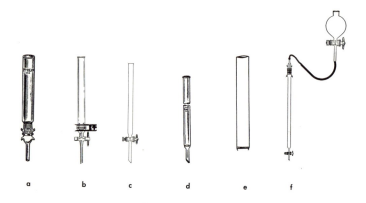

FIG. 4.2. Some types of chromatography tubes. Types (a), (b), and (d) are fitted with fritted glass discs to hold the adsorbent. Type (c) should be fitted with a small plug of glass wool. Type (e) can be fitted with a perforated steel disc or a Teflon plate. Type (f) has a pressure tube attached so that the development can be carried out under the pressure of a long column of solvent.

type (a) is also available with a perforated surface at the joint, to be fitted with a disc of filter paper to retain the adsorbent. Some adsorbents, particularly silica gel, tend to clog a fritted disc and thus retard solvent flow. Placing a piece of filter paper over the disc surface will prevent this problem.

The majority of methods in use today, however, are planned for complete elution of zones from the column and any of the columns (a–e) are useful for his purpose. Type (c), with a plug of glass wool, is useful especially for narrow columns (less than 2 cm) whereas types (a), (b), and (d), with fritted discs, can be used with columns of any diameter. The type (e) column is ordinarily intended to be less than 1 cm in diameter and is fitted with a perforated Teflon or stainless steel plate which in turn supports a disc of filter paper. Types (c) and (e) can readily be prepared with simple glassblowing and machine shop equipment. All of the columns except (e) can be fitted with stopcocks to control flow rate, but the space beneath the column of adsorbent should be minimal to reduce the holdup volume and prevent solute remixing. Column (e) circumvents this problem.

Columns can be fitted with reservoirs at the top, and these can be fitted to pressure sources to speed the flow of the eluting solvent in large columns or columns of finely divided adsorbents. An arrangement as shown in (f) can also be used to control flow rate.

Most of the equipment (a–d) shown in Figure 4.2 is commercially available from a number of the supply houses listed in the Appendix.

The amounts of adsorbent needed for the separation of a given amount of solute vary widely. For preparative separation of some simple mixtures of very different solutes or for crude separations of complex mixtures, adsorption columns are usually run with a solute:adsorbent ratio of 1:20 to 1:50. For complex mixtures or where the method is intended to separate the majority of the components of a simple mixture as pure compounds, lower ratios of 1:200 to 1:2000 are required.

The Adsorbent

The adsorbent and its form are a matter of primary importance in the development of a chromatographic system. Although many substances have been used, specific attention will be given here to the use of two polar adsorbents, alumina and silica gel. These are the most versatile adsorbents available.

Any adsorbent can be prepared and treated to modify its capacity and properties. A number of attempts have been made to control adsorbents and to describe procedures for producing equivalent materials. Such equivalence, however, is somewhat relative. No two brands of an adsorbent are necessarily equivalent even though they may be quite identical in terms of activity grade (see below). The methods of preparation and pretreatment contribute measurably to their adsorption capacity and their ability to resolve various mixtures.

The adsorbent in adsorption chromatography functions as an activated surface that can attract and hold solutes to a given degree. The "activation" of these materials is primarily a matter of the removal of the surface water that is normally present, thus freeing sites for solute adsorption. The degree of this activation has been the subject of much study, and a scale of activity (the Brockmann Index) has been devised for alumina and applied, subsequently, to other adsorbents. The grades of the scale range from I to V where grade I is the most active (containing the least water) and grade V is the least active (containing the most water). The specific details of these scales will be described in the following sections.

The particle size of column adsorbents is generally larger than TLC adsorbents, being in the general range of 80–200-mesh (see Table 4.1). TLC adsorbents will generally pass a 200-mesh screen and are too fine for column work.

Alumina.   Alumina (Al_2O_3) is one of the most widely used adsorbents and it is available in several modifications. Alumina possesses such sites as $Al^{\delta+}$, Al–OH, Al–O$^-$, Al–O$^-$H$^+$ and, depending upon its preparation, also sodium or hydrogen ions. Almost all organic compounds except saturated aliphatic hydrocarbons are adsorbed to the common form, basic alumina. Alumina, however, can be treated with hydrochloric acid to convert it to an acid form or with nitric acid to convert it to a neutral form. Both basic alumina, containing aluminate centers, and acid alumina, containing chloride ions, can function as ion exchangers. Basic alumina will exchange with inorganic and organic cations, and acidic alumina will exchange with inorganic or organic anions. Acidic alumina is largely used for the separation of mixtures of dicarboxylic amino acids and acid peptides, whereas neutral alumina is used for the separation of ketosteroids, glycosides, ketals, lactones, and some esters, and for the dehydration of solvents. Basic alumina has the widest range of application. Highly polar compounds are strongly adsorbed on this material whereas nonpolar compounds (except for unsaturated hydrocarbons) are weakly bound. Acetone should not be used as an eluting solvent with highly active basic aluminas since it will be condensed to diacetonealcohol by an aldol condensation.

Alumina for adsorption is available from a number of firms and in a number of grades and qualities (see Table 4.1). Most laboratory chemical supply companies supply aluminum oxide or alumina in a variety of grades, some of which are intended for the preparation of activated alumina by the investigator. If one is planning to simulate a TLC process, the product that is to be used in the column should be that adsorbent that was used to prepare the TLC product. Thus, if aluminum oxide G was used in TLC the equivalent grade of alumina can be obtained from Merck for use in column adsorption processes.

Basic alumina, grade I, can be purchased as such (Table 4.1), or it can be prepared by heating any available alumina (basic) at 380–400° for three hours, with occasional stirring. Such preparations usually contain some free alkali but this is typical of most preparations sold as basic alumina. If it is desirable to remove the alkaline substances, this product should be boiled repeatedly with distilled water until the washings are neutral and then washed with methanol. Activation, as above, at 200° will again produce an activity grade I product. This material is still to be regarded as basic alumina. Since grade I adsorbents may be too active (causing polymerizations, dehydrations, etc.), grades with lesser activity are generally used.

Alumina of grades II, III, IV, and V can be prepared by adding 3, 6, 10, and 15% water to the grade I adsorbent. In practice this is best achieved

TABLE 4.1

Sources and Properties of Alumina and Silica Gel for Adsorption

Brand	Grades	Particle Size	U.S. Distributor [b]
Alumina			
Woelm [a]	I, basic, pH 10	—	Alupharm
	I, neutral, pH 7.5	—	Chemicals
	I, acidic, pH 4	—	
Bio-Rad [a]	I, basic, pH 10	100–200; < 200-mesh	Bio-Rad
	I, neutral, pH 7	100–200; < 200-mesh	Laboratories
	I, acidic, pH 4	100–200; < 200-mesh	
Fluka [a]	I, basic, pH 9.5	—	—
	I, neutral, pH 7.5	—	
	I, weakly acidic, pH 6	—	
	I, acidic, pH 4.5	—	
Fisher	I, basic	80–200-mesh	Fisher Scientific Co.
	I, neutral	80–200-mesh	
	I, acidic	80–200-mesh	
Merck [a]	I/II, basic	—	Brinkmann
(Darmstadt,	I/II, neutral	—	Instruments, Inc.
Germany)	I/II, acidic	—	
Alcoa	Basic (Cat. #F-20) [c]	80–200-mesh	Aluminum Company of America
Silica Gel			
Woelm [a]	I, pH 7	—	Alupharm
Merck [a]	—	< 0.08 mm	Brinkmann
(Darmstadt,	—	0.02–0.05 mm	
Germany)	—	0.2–0.5 mm	
Fisher	—	28–200-mesh	Fisher
Davison	—	28–200-mesh	W. R. Grace & Co.,
	—	100–200-mesh	Davison Chemical Div.
Mallinckrodt	—	100-mesh	Mallinckrodt Chemical Works

[a] These manufacturers offer adsorbents for both TLC and column chromatography. If the adsorbents for both techniques are derived from the same source, the successful transposition of solvent systems from one technique to the other will be more likely.

[b] Addresses are listed in the Appendix.

[c] Listed as "activated," but must be reactivated before use.

by adding the water to a clean beaker or widemouthed bottle, swirling the container to distribute the water over the walls of the vessel, and adding the adsorbent immediately with the swirling motion being continued. The adsorbent should then be transferred to a powder blender or the distilling flask of a flash evaporator and blended or rotated for at least one hour.

The actual grading of alumina to ascertain its activity on the Brockmann scale can be carried out using a series of azo dyes.[1,2] The dyes (see Table 4.2) are chromatographed on short (1.5 × 10 cm) columns of the adsorbent to be graded, and the resulting positions of the dyes are observed. A comparison of these positions with the data in Table 4.2 will establish the activity.

TABLE 4.2

Position of Dyes on Grades of Alumina Columns [a]

Activity Grade	I	II		III		IV		V
Dye Mixture [b]	1	1	2	2	3	3	4	5
Column position								
Top	a	a	c	c	d	d	e	f
Bottom	b		a		c		d	e
In filtrate		b		a		c		

[a] 1.5 × 10 cm, filled to 5 cm.
[b] Dissolve in pairs (0.02 mg each/10 ml benzene and 40 ml petroleum ether). Pair $1 = a$, p-methoxyazobenzene and b, azobenzene; pair $2 = a$ and c, sudan yellow; pair $3 = c$ and d, sudan red; pair $4 = d$ and e, 4-aminoazobenzene; pair $5 = e$ and f, 4-hydroxyazobenzene. Add 10 ml of test solution and wash with 20 ml of the benzene-petroleum ether mixture.

The above process is somewhat tedious and a much simpler procedure can be used as a routine method for determining the activity of alumina. Proceed by establishing or preparing the five grades of the adsorbent, making sure that the initial product is grade I. Fill a melting-point capillary tube with each grade. Wet the adsorbent at the open end with a drop of benzene. Break the closed end of the tube and immerse the wet end in a 0.5% solution of p-phenylazoaniline in benzene for a moment. Transfer the tube to a vessel containing a shallow layer of benzene and allow migration of the solvent, by capillary action, to a point near the top of the column. Remove the tube and measure the R_f of the dye band for each activity grade. These values are the control for evaluating future lots of the adsorbent. The approximate R_f values for the dye on alumina are as follows: grade I, 0.0; grade II, 0.13; grade III, 0.25; grade IV, 0.45; and grade V, 0.55.

Neutral alumina can be purchased as such (see Table 4.1) or it can be

[1] H. Brockmann and H. Schodder, Ber. **74**, 73 (1941).
[2] H. Brockmann, Angew. Chem. **59**, 199 (1947).

prepared [3] in the following manner. Normal activated alumina is suspended in water and boiled. The supernatant liquid is made just acid to litmus with dilute nitric acid. The boiling process is continued and sufficient nitric acid is added so that the boiling solution will remain acidic for ten minutes after the last addition of acid. The adsorbent is then filtered and washed with water until the washings are neutral. It is boiled with methanol, filtered and dried at 160–200° under reduced pressure (10 mm) for 12–16 hours. The resulting material is presumably activity grade I. It can be deactivated by the addition of water to the various other less active grades.

Acidic alumina can be purchased as such (see Table 4.1) or it can be prepared in the following manner.[4] Normal activated alumina is suspended in three to four times its volume of $1N$ hydrochloric acid and stirred for 10–15 minutes. The supernatant and very small particles are decanted and the process is repeated several times. The alumina is then collected by filtration on a sintered glass Büchner funnel and slowly washed with water until the washings are only slightly acidic to litmus. It is then dried at 100°. The resulting material is not very active and functions primarily as an anion exchanger in the chloride form.

Silica Gel.   Silica gel (SiO_2), or silicic acid, like alumina is a commonly used adsorbent and might be regarded as the most versatile of all adsorbents. Although the terms silica gel and silicic acid are used interchangeably they are, in fact, modifications of the same material.[5] Silica gel can be used with all solvents, but it shows hydrogen-bonding capacity with some solutes and solvents when water is present. This bonding character together with the fact that it will swell and thus slow solvent flow in the presence of water, methanol, and ethanol causes some limitation to its general use.

Activity grade I silica gel can ordinarily be prepared by heating it at 150–160° with occasional stirring, for 3–4 hours. Although highly active grades were made for many years by heating at 300° or higher there is evidence for irreversible degradation when silica gel is heated above 170°. Grade I silica gel is an anhydrous product; grades II–V are made by adding water to a concentration of 10, 12, 15, and 20% respectively. The capillary-tube standardization process, described on p. 92 under alumina, can be used for the determination of the activity of silica gel. For example, activity grade I has an R_f value of 0.0 and grade III has an R_f of approximately 0.65.

[3] T. Reichstein and C. W. Shoppee, *Discussions Faraday Soc.* **7**, 305 (1949). The process has been modified in I. E. Bush, "The Chromatography of Steroids," Pergamon Press, Inc., New York and London, 1961, p. 352.

[4] T. Wieland, *Hoppe-Seyl. Z. Physiol. Chem.* **273**, 24 (1942), as recorded in I. E. Bush "The Chromatography of Steroids," Pergamon Press, Inc., New York and London, 1961, p. 352.

[5] J. J. Wren, *J. Chromatog.* **4**, 173 (1960).

Preparation of the Column

A properly prepared column is essential for the resolution of mixtures. Adsorption columns may be prepared either dry-packed or wet-packed. In the former case the adsorbent is introduced into the column in small quantities and each quantity is compressed by gentle packing with a plunger. Such plungers may be a glass rod flattened at the end, a rod with a stopper mounted on the end, or a wooden dowel. After packing, the column is washed with the solvent to be used initially in the elution process.

In wet-packing, the adsorbent is made into a slurry with the solvent and poured into the column to fill it (see Figure 1.6). During the settling process excess solvent is drained and an additional quantity of the slurry is added. This process is continued until the desired column height is obtained. The solvent may be one high in the eluotropic series, for example, benzene, or it may be the initial solvent to be used in the elution. If a solvent is used which is different from the initial eluting solvent, the column must be washed with the eluting solvent prior to its use in a chromatographic process.

Alternatively, a wet-packed column may be prepared by introducing the solvent into the tube and then adding the adsorbent in a fine stream, through a funnel, and allowing the adsorbent to settle while the tube is gently tapped with a plastic or wooden rod. The excess solvent, if any, is removed as above. If this operation is conducted without interruption, an excellent column is readily obtained.

In each case excess solvent is drained from the column to provide a continuous column of adsorbent and solvent. The top of the column is covered with a circle of filter paper and some glass wool, glass beads, or washed sand to prevent any disruption of the top surface of the column during further solvent addition.

Solute Introduction and Elution

The solute mixture, dissolved in a minimum volume of the solvent to be used for elution, is introduced into the column and permitted to percolate into the adsorbent (see Figures 1.8 and 1.9). When the solution has disappeared beneath the top of the adsorbent, the eluting solvent is added. Should the mixture be composed of components that are poorly soluble in the solvent, the mixture should be dissolved in a polar solvent and this solution should be added to a small quantity of the adsorbent in an evaporating dish. The solvent should be evaporated and the dry powder added to the top of the column. The eluting solvent is then added in the manner described above.

Flow rate in adsorption chromatography may be more rapid than for other chromatographic processes and is generally determined by the solute-solvent-adsorbent characteristics and the column size. There are no rigid rules, but ordinarily the rate is 5 ml/cm^2 (cross-section area)/hr or more. Slow flow rates usually produce better elution patterns than rapid flow rates. Velocity of flow may be changed during the course of development, but in no case should the elution process be stopped for any prolonged period of time.

It is important that the solvents used in elution are of absolute purity, and this requirement may demand distillation or percolation through an adsorbent prior to use.

If there is evidence of tailing of bands in the effluent it is imperative that a change of solvent polarity be made to provide for more rapid elution. Evidence for tailing is obtained by some quantitation process applied to fractions of the effluent (see next section). A solvent change in adsorption chromatography may be a complete change of solvent, a stepwise addition of a second solvent, or the establishment of a gradient for a pair of solvents.

A complete substitution of the eluting solvent by another can be done provided the change does not cause a change in the order of solute displacement. It is thus necessary to proceed down the eluotropic series (see Table 2.1) in a stepwise manner rather than to make a broad change in solvent polarity. Such stepwise substitutions are usually made in systems involving highly complex mixtures, and usually each solvent is used in the process until the column is essentially stripped of the solute material that it readily elutes. A more profitable experimental method, in terms of both time and material, is to establish a pattern of stepwise changes involving increasing concentrations of the second solvent. Thus column elution is begun with a pure solvent, and after a given volume has been added—usually determined by the analytical profile of the effluent—the second solvent is added to a concentration of 0.1–5%. The concentration is determined by the polarity differences of the solvent pair and by the extent of tailing. If the polarity difference is small or if tailing is severe, concentrations greater than 5% might be used. After a given volume of this solvent has been used, a change is made to a greater dilution and this process may be continued until the second solvent is the eluting agent. A third solvent can then be added in a similar pattern. It is possible, in this manner, to completely elute the solutes from the column and achieve an excellent fractionation. Such techniques can be conducted on small columns for exploratory purposes. In this connection, it should be noted that the polarity changes produced by adding one solvent to another are not linear, but logarithmic (see Figure 2.1).

At the end of any chromatogram, the column should be stripped with very polar solvents so that no solutes will be lost. This can be done with a 5% solution of acetic acid in methanol or, sometimes, with methanol alone.

Apparatus (see Figure 4.3) for the development of an exponential or a linear solvent gradient can be constructed for both adsorption and partition columns. Of these, the linear gradient system is usually the most effective procedure for use as a general method. In adsorption processes the second and more polar eluting solvent is in the reservoir (a).

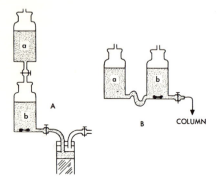

FIG. 4.3. Two designs of apparatus which will provide a gradient solvent system for column development. Design A will give an exponential gradient and design B will provide a linear gradient. A stirring bar is in vessel b to provide efficient solvent mixing.

The major problem in developing a gradient elution system is that of selecting a system which will give the proper gradient curvature. The linear gradient apparatus lends itself to a control of the mixing pattern of the solvents in (a) and (b). By inserting a solid core (this core can be of various sizes) into vessel (a) the volume of solvent in (a) that moves into vessel (b) to mix with the solvent in (b) can be controlled. Since the height of the solvent in each of the two vessels remains at an equal level, the size of the core used to displace the solvent will control the gradient curve.

Detection of Resolved Substances

Most substances resolved by chromatographic processes are colorless. Thus, a system must be designed to collect fractions of the effluent. In general, fractions are divided according to volume or time intervals. A number of commercial fraction collectors are available which permit collections according to volume, by drop counting, or by time intervals. Those collectors which make use of a siphon apparatus for volume collection usually permit some mixing of the effluent in successive fractions and are less desirable for difficult separations. Figure 4.4 illustrates a turntable-type collector whose movement is actuated by a device that counts drops of the effluent.

Monitoring of the fractions is usually accomplished either directly by physical or chemical detection methods or somewhat indirectly by TLC analysis. Since chromatography deals with the whole of organic and inorganic matter, it is not possible to present a simple explanation of methods and problems of fraction assay. Physical properties such as color, fluorescence, optical activity, light refraction, pH, and radioactivity are widely used methods, but there are many other properties that may be used. Chemical methods usually involve the application of some reagent that produces some readily discernible reaction such as color change or color development. The selection of the reagent depends upon the chemical properties of

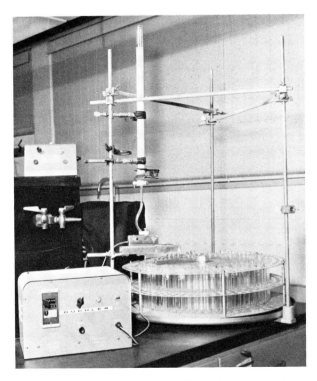

FIG. 4.4. A commercial fraction collector for use in column chromatography. (Reproduced through the courtesy of Buchler Instruments.)

the component(s) of the mixture. Reagents or physical methods of analysis can be applied directly to the various fractions, but interference by the solvent must be considered in each case. The individual fractions used for the test may, however, be dried (sometimes weighed) and the test applied to a portion of the residue or a solution of such a portion. An example of the effluent profile is shown in Figure 4.5A.

The effluent may also be evaluated by biological methods. Biological effects on animals or tests for enzymatic activity are widely used methods, but other *in vitro* tests may be used to monitor extracts containing solutes which possess physiological activity. Compounds which possess antibiotic effect or those which affect growth or show nutritional effects can be evaluated in *in vivo* systems employing microorganisms or isolated animal tissue.

One of the simple general methods of monitoring effluent fractions is by TLC analysis (see Figure 4.5B). With the use of a specific reagent for the

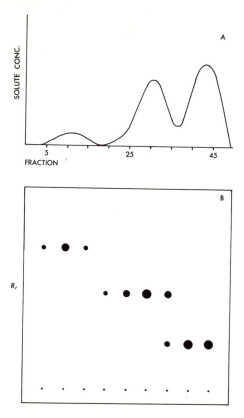

FIG. 4.5. A shows an idealized curve obtained by analyzing fractions (0–45) from a column effluent. B shows the TLC picture which would correspond to the curve. Every fifth fraction was analyzed.

compound in question and a general reagent, such as sulfuric acid, to reveal organic matter generally, it is possible to develop a profile of elution that gives a rapid and clear picture of the effluent composition. It is one of the best methods, also, to reveal fraction tailing, thus providing the necessary evidence for a solvent change in the elution pattern.

Most of the methods used are qualitative in nature, but fractions can be evaluated quantitatively by weighing and by the physical, chemical, and biological methods described above.

Systematic analysis of the effluent need be concerned, at the beginning, only with fractions taken at various intervals—for example, every fifth or tenth fraction. The profile thus obtained will indicate the extent that intermediate fractions need to be analyzed to obtain a complete profile. Fractions of equivalent composition should be combined. In some cases these will be resubmitted to another chromatographic process for complete fractionation.

In all cases of monitoring the effluent, it is important to realize the limitations of the method used for revelation. It has already been mentioned that the solvent may cause interference, but equally important are those phenomena associated with the limits of detectability and the effect of other solutes. The investigator must be knowledgeable of the sensitivity limits of the procedure he is using, and it is advisable that he not rely upon one reagent alone for the initial revelation of a specific component.

Product Isolation

The column fractions that contain identical components are combined and the solvents are evaporated, preferably under vacuum. Frequently, the products will require recrystallization to yield samples with optimum purity.

Column Assay by Extrusion

On occasion, column chromatography is carried out so that none of the solute components are actually washed out of the adsorbent (see pp. 6 and 88). In this case the adsorbent is extruded or pushed out of the column to give a core that will contain bands, each containing a solute component. This core is analyzed by U.V. light if the solutes fluoresce or if the adsorbent contains a phosphor which will reveal them. Alternatively, the core can be treated on one side with some visualization reagent (applied with a brush or dropper). The band location is marked and the adsorbent containing the reagent is scraped off and discarded. The solute components are then washed or eluted from the adsorbent bands and isolated.

4.3. COLUMN PARTITION CHROMATOGRAPHY

In column partition chromatography a solid support is made to retain a stationary liquid phase and resolution of the solute mixture is obtained by downward flow of a mobile phase. It is thus a liquid-liquid distribution process. In conventional practice, the more polar of the solvent pair is held by the support (**normal partition**). In **reversed phase** chromatography the position of the solvent pair is reversed.

In general, column partition chromatography is more difficult to carry out than column adsorption chromatography. The capacity of partition systems is low, and it is not easy to impregnate the supports and obtain an equilibrated system. However, they do have a great resolving power and must be used for very polar solutes.

Columns

Columns for partition chromatography are of the same design as those for use in adsorption chromatography (see Figure 4.2), but the width:length ratio is ordinarily at least 1:20. Since column length determines the number of partitions that can occur as the mobile phase moves past the stationary phase, column length must increase as solute mixtures become increasingly complex, or separations will not be complete. Column length is also determined, in part, by the particle size of the support. Longer columns must be used if the particle size is relatively large, whereas shorter columns can be used if the particle size is smaller.

The ratio of solute to support must be lower than the comparable ratios used in adsorption work. These will vary widely, but are frequently in the range of 1:100 to 1:2000.

The Support

The support for the stationary phase should be selected for its capacity to hold that phase in relatively large amount, and it should not interact with the phases or the solutes in any manner. The support is ordinarily finely divided to provide a large area for the stationary phase. The silicates and powdered cellulose are the most widely used support materials, but starch and a few other materials have been found useful. The silicates, kieselguhr and silica gel, offer the least complications, and the beginner is advised to select one of these in preliminary studies. Each of these and cellulose are available in a variety of grades (see Table 4.3) useful in a broad variety of partition systems, and each can be used to support a polar or a nonpolar stationary phase.

Silica gel.   Silica gel for column partition chromatography is usually graded to precise particle size limits. This is necessary to permit proper packing or preparation of the column and is a determining factor in flow rate of the mobile phase. The sizes most commonly used range within the scale of 28–200-mesh, and products within this range are capable of retaining 50–75% of their weight of aqueous stationary phase for normal column partition chromatography.

Silica gel, prepared for adsorption processes, can be used for partition processes provided it is deactivated. This is normally accomplished by impregnating it with the stationary phase to be used in the chromatogram (p. 104).

Commercial grades of silica gel usually contain appreciable quantities

(10–30%) of water and may contain impurities. If analytical data are not available, the product should be cleaned by washing in two volumes of a mixture of equal parts of hydrochloric acid and water for 6–12 hours. After the liquid is decanted, the silica gel should be thoroughly washed with distilled water and finally with methanol or ethanol. Drying at 110–120° will provide a sample that is suitable for impregnation with a stationary phase.

If such a product is a mixture of a broad range of particle sizes, it should be sieved to a proper mesh size. Two sieves are required to prepare a product that will fall within minimum and maximum size particles; for example, 100-mesh and 200-mesh.

TABLE 4.3

Sources and Properties of Support Materials for
Column Partition Chromatography

Brand	Properties	U.S. Distributor
Kieselguhr (All Analytical Grade)		
Filter-Cel	surface area: 16 m²/g	Johns-Manville Corp.
Super-Cel	surface area: 5 m²/g	Johns-Manville Corp.
Hyflo Super-Cel	surface area: 2 m²/g	Johns-Manville Corp.
Celite 503	surface area: 1.5 m²/g	Johns-Manville Corp.
Celite 545	surface area: 1 m²/g	Johns-Manville Corp.
Silica Gel		
Davison 12	28–200-mesh	W. R. Grace & Co., Davison Chemical Div.
922	200–325-mesh	(Fisher Scientific Co.)
63	325-mesh	(Fisher Scientific Co.)
35	12–42-mesh	(Fisher Scientific Co.)
Bio-Sil HA	> 325-mesh	Bio-Rad Laboratories
Bio-Sil BH	100–200-mesh	Bio-Rad Laboratories
Bio-Sil A	100–200- and 200–325-mesh; highly purified	Bio-Rad Laboratories
Merck 7734 (Darmstadt, Germany)	68–320-mesh	Brinkmann Instruments, Inc.
Merck 7754	68–320-mesh; especially purified	Brinkmann Instruments, Inc.
Merck 7733	35–68-mesh	Brinkmann Instruments, Inc.
Cellulose		
Cellulose Powder, MN2100	acid-washed; fat-free	Macherey, Nagel & Co. (Brinkmann Instruments, Inc.)
Cellulose Powder, S & S 389	ashless	Carl Schleicher & Schuell Co.
Cellulose Powder, Whatman	ashless	Reeve Angel and Co.

Because silica gel possesses adsorptive properties for some solutes, kieselguhr is a more ideal support for use in preliminary studies.

Kieselguhr.   Most commercial grades of diatomaceous earth are composed of extremely fine particles and these products usually contain soluble impurities. Treatment (cleaning and sieving), as described under silica gel will provide a sample suitable for use in partition systems. Those products marketed as analytical grade quality (see Table 4.3) can be used without treatment. These products are available in graded mesh sizes. Kieselguhr is widely used in both normal and reversed phase systems and it is usually the support of choice for the latter type of system.

Cellulose.   Powdered cellulose was extensively used in column processes when paper chromatography was in vogue because of the direct application of solvent systems derived from paper strip systems to preparative column systems. Since paper partition chromatography is no longer the most popular method for preliminary studies, cellulose columns are only infrequently used as a general preparative method.

Various grades of cellulose powder are available, but those described as ashless should ordinarily be used. Cellulose columns are difficult to prepare to provide a uniform structure, and cellulose is further limited as a general support because the ratio of mobile phase to stationary phase is ordinarily large.

The Solvent System

The solvent system, in its simplest form, is an equilibrated pair of immiscible liquids for use as the stationary phase and the moving phase. Equilibration is obtained by thoroughly mixing the solvents in a separatory funnel before use. The individual phases are then used to *prepare* and to *develop* the column. To prepare the column the support material is mixed with the equilibrated phase that is to be held stationary. In normal partition chromatography this is the more polar mixture. The other phase is then used to develop the chromatogram. The solute mixture, having been introduced to the top of the column, distributes itself between the mobile and stationary phase, and the relative position of any one solute, in a flowing chromatogram, is determined largely by its relative solubility in the two phases. Those substances which have an affinity for the stationary phase solvent move slowly whereas those which have a greater affinity for the mobile phase solvent move more rapidly through the support. These factors are discussed more completely in Chapter 2.

It is important that the solvents used are pure since impurities will cause significant changes in solute distribution. It is further important that the temperature of equilibration and that for use in the chromatographic process be the same to prevent changes in mutual solubility of the solvent pair.

The solvent pair of such a biphasic system is selected according to the solubility characteristics of the solute mixture. The solute mixture must distribute itself between the two phases. A ratio of solubility of 1:100 (solubility in the moving phase:solubility in the stationary phase) for the major component of the solute mixture is an ideal value. However, such solubility information is frequently not available. Water is the most common stationary phase solvent and *n*-butanol, benzyl alcohol, phenol, ethyl acetate, chloroform, and benzene are commonly used mobile phase solvents (see Table 2.4). Aqueous acids or bases and aqueous buffers, as the stationary phase, together with the above list of organic solvents constitute other typical biphasic systems. Biphasic systems are also composed of a mixture of three or more organic solvents or two or more organic solvents and water to provide workable systems. Hydrophilic solvents other than water may be used. These include the lower-molecular-weight members of the alcohol and glycol series and formamide. They are particularly useful for the separation of solutes of medium or low polarity.

All of these systems can be reversed in the process known as reversed phase chromatography, although special solvent systems are usually devised for the purpose. These include the use of mineral oil, silicone oil, or other hydrophobic materials as the stationary phase. In these cases the support is impregnated with a solution of the hydrophobic material, for example mineral oil in acetone. After this mixture (the support and the mineral oil) has been equilibrated with the polar mobile liquid, for example acetone-water, it is introduced into the column and used in a partition system, with the acetone-water solution as the mobile phase. The reversed phase systems cited in Table 2.4 will illustrate the variations that have been used.

Solutes, whose solubility is related to the ionization character of the molecule, are readily separated in partition systems using a gradient elution technique. This method is widely applied to the separation of mixtures of organic acids. In this case the stationary phase is usually a strong acid, such as hydrochloric acid or sulfuric acid, diluted with water, and it is of such strength that it will depress the ionization of the solute components. The gradient is obtained by the addition of water-saturated butanol to water-saturated benzene, for example, to provide a constant increase in the polarity of the mobile phase. It is impossible, under these conditions, to maintain a perfect equilibration between the two phases, but the method has been shown to be functional in a variety of experiments.

Preparation of Partition Columns

Columns for partition chromatography are more difficult to prepare than columns for adsorption chromatography. Because of the differences in the nature of the support material and the system used, a number of methods have evolved for the preparation of partition columns.

With kieselguhr and silica gel, the stationary phase in normal partition systems is ordinarily mixed directly with the support. From 0.5 to 1 part, by weight, of the equilibrated stationary phase is added to the support and intimately mixed by grinding with a pestle in a mortar or other vessel, or in a powder blender if large amounts are to be prepared. The powder thus prepared is not wet, but it is also not free-flowing. Silica gel will ordinarily hold more water than will kieselguhr, and fine grades of silica gel will hold more than will those that are of large mesh size. This powder is then shaken with an excess of the previously equilibrated mobile phase to form a slurry and this mixture is poured into the column. Columns of silica gel will form readily by gravity, but a kieselguhr column must be packed. For those columns formed by gravity the excess solvent is allowed to run out slowly and the slurry is added in portions to fill the column to the desired height. Packing of kieselguhr columns may be accomplished by tamping with a plunger on the end of a rod or with the use of a perforated ($\frac{1}{32}$–$\frac{1}{16}$ in. holes) steel disc mounted on the end of a rod. In each case the diameter of the plunger or disc should be just slightly less than the inside diameter of the tube. The plunger or disc is used first to remove air bubbles from the slurry by several rapid up-and-down motions and then to press small sections of the column by a slow downward motion of the device. This method also requires that successive small quantities of the slurry be added to the column.

Cellulose columns are best prepared by the gravity technique using a slurry of the powder in acetone. Excess acetone is drained from the column and the column is washed with the equilibrated mobile phase (sometimes initially with an excess of water) until equilibration is obtained. Evidence of equilibration is noted by constancy in pH, refractive index, or some other criterion in the effluent. Columns prepared under these conditions will absorb water from the mobile phase and will swell to form a uniform column.

If a column is to be prepared for reversed phase chromatography, the support (usually kieselguhr or cellulose) must be dried to remove water. The dry powder is then usually made hydrophobic by exposing it for 12–16 hours to the vapor of dichlorodimethysilane in a desiccator. This powder should be washed with water (it will float) and methanol until the filtrate is not acid to bromothymol blue. After drying at 110° it may be stored for later use. The dry powder, either with or without the silane treatment, is mixed with the less polar phase of the equilibrated solvent mixture and then suspended in the more polar phase for preparation of the slurry and the column as described above.

Solute Introduction and Elution

This operation is fundamentally like that used in adsorption columns. The solute mixture is dissolved in a minimum volume of equilibrated mobile phase and this solution is added to the top of the column. Alternatively, if this volume is large because of poor solubility, the solute mixture can be dissolved in the stationary phase and this solution added to dry support material (ca. 0.3 ml/g). This powder is then placed at the top of the column and the mobile phase can be added to develop the column. The top of the support column should always be protected with a filter paper disc.

Stepwise changes in the eluting solvent can be made, but only after one or several bands of solute have been eluted and it is known that this solvent will not resolve the remainder of the solute mixture. The new solvent is ordinarily equilibrated with the stationary phase liquid before use. An example of such a system for fatty acid separation [6] and for amino acid separation [7] should be consulted to provide a greater insight into the technique.

Gradient elution processes (see p. 96) in column partition chromatography make use of acids or bases to control the ionization and thus the solubility of solutes in the stationary phase. Mobility of the solutes is obtained by use of a gradient in which the polarity of the mobile phase is increased. Such control tends to create sharp, narrow bands of ionizable solutes. It is difficult to maintain an equilibrated system in such processes, but it is essential that the mobile phase components be equilibrated with the stationary phase solvent. An example of a system for the quantitative separation of a wide range of organic acids in plant material or for pure acids [8,9] will serve as excellent models for the design of an experiment.

Evaluation of the elution process and product isolation are conducted by methods that have been described in the previous section.

[6] G. A. Howard and A. J. P. Martin, *Biochem. J.* **46**, 532 (1950).

[7] W. H. Stein and S. Moore, *J. Biol. Chem.* **190**, 79 (1949).

[8] J. R. Lessard and P. McDonald, *J. Sci. Fd. Agric.* **17**, 257 (1966).

[9] G. G. Freeman, *J. Chromatog.* **28**, 338 (1967).

5

Gas Liquid
Chromatography

5.1. INTRODUCTION

Gas liquid chromatography, or GLC, is the most highly instrumented of the techniques discussed in this book. At first glance, GLC equipment would appear to be quite complicated. This is not necessarily so, however, as we will try to show. The basic steps of the method and the underlying principles have been discussed briefly in Chapter 1 and are illustrated in Figures 1.10 and 1.11.

GLC can be used on any mixture of compounds in which at least one of the components has an appreciable vapor pressure at the proposed temperature of the separation. However, to obtain a complete analysis, all of the components should be somewhat volatile at the temperature used. This volatility criterion now includes most organic and many inorganic materials, for GLC can be carried out up to 400°.

The separation of a mixture by GLC is rapid in comparison to the classical techniques of fractional distillation, fractional crystallization, or column chromatography, since the length of time required for a single separation is only 20–60 minutes. It must be added, however, that the classical techniques will usually, but not always, permit a larger amount of material to be separated. The GLC separation, as was mentioned in Chapter 1, has the added advantage of resolving mixtures containing up to 50 or more components. Furthermore, the resulting chromatogram will give information about the amounts of the components as well as their identities.

Another timesaving attribute of GLC is that the chromatographic column is usually continuously regenerated. Thus a GLC column will last almost indefinitely with normal care and use. If a large amount of nonvolatile

106

material is placed on the column either in one operation or over a long period of time, the efficiency of the stationary phase would, however, be expected to drop.

GLC has a particular feature that is only partially present in other forms of chromatography, for its stationary phases are almost infinite in number. The liquid coating on the solid support in the column can be almost any kind of chemical compound. Also, and this fact is quite important, a liquid can be chosen which will separate a mixture either in terms of the boiling points of the components or in terms of their chemical nature.

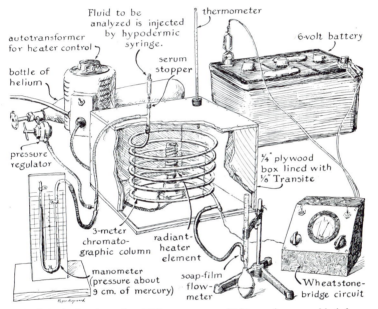

FIG. 5.1. A drawing of a GLC apparatus which can be assembled from available materials. The original paper should be consulted for further details. (From "The Amateur Scientist," by C. L. Stong. Copyright © June, 1966 by Scientific American, Inc. All rights reserved.)

GLC is probably the most versatile chromatographic method because of the wide variations in stationary phase. However, its equipment can be the most expensive and complex. Even the simplest homemade apparatus (a recent one is depicted in Figure 5.1)[1] involves quite a number of different items, some of which are moderately expensive. When the apparatus in Figure 5.1 is examined, however, it is apparent that it can be built by high school students and others interested in science projects.

Because of time and ability problems, many people prefer to purchase

[1] Additional references to simplified apparatus are given in the Appendix.

their GLC apparatus. All have the same basic parts as the one in Figure 5.1. Two chromatographs are pictured in Figure 5.2.

Operational Directions

Fortunately, no matter how simple or complex a GLC apparatus may be, all are operated in basically the same manner. The operation will be described as a series of steps, and the subsequent sections will give more information about each step.

(1) The apparatus is inspected. This is done to ascertain that the correct column is in the apparatus and that the fittings are tight.

FIG. 5.2. Commercial gas chromatography apparatus. The apparatus on the left is a Varian Aerograph HYFI, Model 1200, and the apparatus on the right is an Aerograph HYFI, Model 600. The recorder in the center is a Honeywell recorder equipped with a Disc Integrator.

(2) The gas flow through the column is started. This is done by opening the main valve on the gas tank and then turning the secondary (diaphragm) valve to about 5 lbs. This will permit a slow flow of gas, 5–10 ml/min, to pass through the column and protect it and the detector against oxidative breakdown. See p. 115 for more details on the gas flow.

(3) The column is heated to the desired temperature. This is done rapidly by turning the variable voltage transformer, which controls the heating coils in the oven or furnace, to about 90 volts. When the temperature is 10–15° below the desired temperature, the transformer is turned to that voltage (10–50 V) which will keep adding enough heat to balance the heat loss. An apparatus with a direct dial temperature control is easier to operate, but also much more expensive. See p. 123 for information on choosing the temperature.

(4) If there are separate heaters for the injector and detector, they are turned on. Their temperatures should be about 10–15° higher than the column temperature. It is also wise to check the septum in the injector port to make sure it is not defective. See p. 118 for other information.

(5) The flow of carrier gas through the column is regulated to about 60 ml/min when a ¼ in. column is used. A soap bubble flowmeter is usually utilized to measure the flow rate. If the apparatus is not equipped with one, see p. 114 for construction details.

(6) The current to the detector is turned on. The amperage is regulated to the proper amount depending on the detector, and is usually 150–200 mamps for a hot-wire type. After the detector chamber is warm (2–3 min), the electrical circuit is balanced. See p. 124 for specific information on the various detectors.

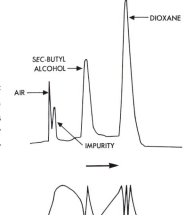

FIG. 5.3. A recording of a gas chromatogram of a 1:2 mixture of sec-butyl alcohol and dioxane on a nonyl phthalate column. The peak which is present in addition to the air peak and the major peaks suggest at least one impurity. The continuous line underneath is the integrator output.

(7) The sample is injected. A small amount (10 μl–1 ml) of the liquid is taken up into a syringe and the sample is placed on the column by forcing the needle through the rubber injection-port septum to its fullest extent and then rapidly forcing the liquid out of the syringe. The syringe should then be removed and immediately cleaned. Page 115 affords more details.

(8) The peaks are recorded to produce the chromatogram. This can be done manually or on a strip chart recorder.

Figure 5.3 shows the chromatogram which resulted when a mixture of sec-butyl alcohol and dioxane was separated by the GLC apparatus in Figure 5.2. The output was recorded on a strip chart recorder. The number of peaks indicates that at least three components (one impurity) were present. The particular recorder used has an attachment (an integrator) which measures the areas under each peak and records this information in a line below the peaks. Thus, the chromatogram has information that can be used to obtain quantitative information (Section 5.5) as well as qualitative (Section 5.4). This specific separation will be used to illustrate a number of points in GLC. Thus, it appears in Figures 5.4, 5.5, 5.10, 5.12, and 5.13.   *P 130 , P.134,*

Conl GLC is not without problems and some of these are: determination of the correct temperature for the separation, choice of the correct stationary phase, **bleeding** of the column, partial resolution of the components, the insensitivity of certain substances to detection, etc. These problems and their solutions, in addition to more details on the parts of a system, are the subject of the subsequent sections.

5.2. CHOICE OF A SYSTEM

There are three major variables in GLC. These are, in order of increasing complexity: the carrier gas, the temperature of the separation, and the nature of the stationary phase.

The Carrier Gas

The force which causes a substance to move through a gas chromatograph is primarily the volatility inherent in the substance itself. The exact nature of the carrier gas is secondary as far as separations are concerned, and the specific gases used will depend primarily on the detector used in the system. Four major detectors are in common use: the thermal conductivity detectors (hot wire or thermistor), the flame ionization detectors, ionization detectors (radioactive source), and the electron capture detectors. The first of these, the thermal conductivity detectors, are most frequently used in simple systems.

Nitrogen, helium, argon, and sometimes hydrogen are used as carrier gases since they are quite unreactive and can be purchased pure and dry in large-volume, high-pressure tanks. Although helium will give the greatest sensitivity to a thermal conductivity detector, it is inferior to nitrogen in that more lateral flow and mixing occur with the less dense helium. Despite this disadvantage, helium is generally used. Ionization detectors involving a radioactive source require argon or helium. Flame ionization detectors can be used with any inert gas; however, nitrogen is generally used. Electron capture detectors require nitrogen or argon.

Compressed air, which is readily available, cannot be used because oxygen will oxidize the liquid stationary phase, the detector, and the materials in the mixture undergoing separation. However, in a simple apparatus (thermal conductivity detector) it is possible to use natural gas (propane or butane from the laboratory gas jet) after taking the normal precautions to preclude any explosion or fire.

Temperature

The temperature at which a separation is carried out is very much a matter of experimentation. A good initial choice is a few degrees below the boiling point of the major component of the mixture. If all of the materials in the mixture have the same or nearly the same boiling point, the separation will be completely dependent upon the nature of the stationary phase.

The Stationary Phase

The correct column for any separation depends primarily on the type and amount of stationary phase, but there are minor influences from the diameter, length, and type of tubing, and the type and mesh of the solid support. The details on the tubing and solid support will be presented in the succeeding sections; here we shall consider the types of stationary phase.

"Like dissolves like" is the major criterion used in choosing a stationary phase. The word "dissolves" is appropriate, for the separation depends upon the components in the sample being dissolved in the stationary liquid to different degrees (see Chapter 1). If one or more components in a mixture are insoluble in the liquid phase used, the component or components will rapidly pass through the column with no separation occurring. This absence of solution is true of air, which is either present in all samples or is placed in all samples intentionally. Air is both insoluble in normal stationary phases and extremely volatile when compared to the materials being separated. Thus, it emerges first and indicates the minimum time for passage through the column (see Figure 5.3).

The separation of a mixture of materials that are of the same chemical nature is most often effected by a boiling point separation. This requires that the stationary phase be similar in chemical nature to the compounds being separated in order to achieve satisfactory separation, i.e., a high-boiling aliphatic hydrocarbon for aliphatic hydrocarbons. The aliphatic hydrocarbon stationary phase would also be useful for substances close in structure and polarity to an aliphatic hydrocarbon, such as an aliphatic ether or an alkyl halide. In a similar manner, highly polar materials, such as alcohols and amines, require a highly polar stationary phase for their separation. Some materials, such as the phthalate esters, have been found useful for the separation of many different kinds of substances.

The "like dissolves like" criterion is often applied to parts of mixtures, since two materials of about the same boiling point and different polar nature can be separated on either a nonpolar or a highly polar stationary phase. An example is the separation of dioxane (b.p. 102°) and *sec*-butyl alcohol (b.p. 100°) on a highly polar polyethyleneglycol (Carbowax) column. This is illustrated in Figure 5.4, which also shows the use of nonyl phthalate and other columns at the same temperature and gas flow for the separation of dioxane and *sec*-butyl alcohol. A Versamid (a high-molecular-weight polyamide) and two different silicone oil columns did not give a separation of the mixture.

Some stationary phases contain complexing agents. This involves the complexing (a chemical interaction and not just dissolving) of one of the components of a mixture by a material in the stationary phase. The best example is the separation of an olefinic hydrocarbon from a saturated hydrocarbon of the same boiling point by the complexing of the olefin with

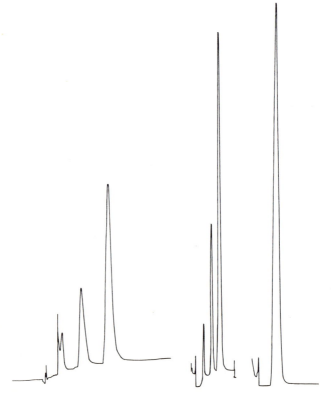

FIG. 5.4. Three chromatographic curves which show the influence of the choice of stationary phase on the separation of the *sec*-butyl alcohol–dioxane mixture used before (Fig. 5.3). The curve on the left was measured on a nonyl phthalate column; the middle one was measured on a 5 ft Carbowax column, and the right curve was measured on a 5 ft Versamide column. All three columns are commercially available, but can also be fabricated.

silver ions present in the stationary phase. This is similar to the use of silver ion in thin layers on p. 56.

Table 5.1 gives a recommendation for the first choice of a liquid phase and its maximum operating temperature when a particular type of compound constitutes the mixture that must be separated. This table lists as few stationary phases as possible, but should be enough to permit the purchase or preparation of a minimum of four columns for general GLC analysis. Most GLC suppliers provide purified stationary phases. The columns indicated are not as specific as might be thought, for either a silicone oil or alkyl phthalate column can be used for aromatic hydrocarbons, and a paraffin oil or silicone oil column will resolve aliphatic hydrocarbons and ethers. Although it is not stressed in Table 5.1, the silicone oil column has proven to be one of the most versatile available.

TABLE 5.1

First Choice of a Column Stationary Phase [a]

Type of Sample	Stationary Phase (Max. Temp.)
Aliphatic hydrocarbon	Paraffin oil (Squalene, 300°) (Apiezon)
Aromatic hydrocarbon	Dinonyl phthalate (175°)
Alcohol	Dinonyl phthalate (175°)
Polyol	Polyethyleneglycol (Carbowaxes, 125–200°)
Aldehyde or ketone	Paraffin oil (Squalene, 300°)
Acid	Silicone oil (DC 200, 250°)
Ether	Dinonyl phthalate (175°)
Ester	Dinonyl phthalate (175°)
Amine	Polyethyleneglycol (Carbowax, 125°)
Alkyl halide	Dinonyl phthalate (175°)
Sulfur compound	Silicone oil (DC 200, 250°)
Gases	Molecular sieve

[a] Both prepared columns and purified phases can be purchased from most GLC suppliers; see Appendix.

A number of other stationary phases should be mentioned, for they have proven to be valuable for certain separations. These include the DEGS (diethyleneglycol succinate) column for highly polar materials and isomeric materials; SE-30 (silicone grease) for acids, hydrocarbons, and many complex natural products; and FFAP (free fatty acid phase, a modified Carbowax [2]) for alcohols, ketones, esters, nitriles, and sulfur compounds. Most GLC catalogs [3] give an extensive amount of information on many special stationary phases to use with specific mixtures, and will also indicate the maximum temperatures for each stationary phase.

[2] A Varian Aerograph material.
[3] See Appendix for names.

5.3. THE SYSTEM

Commercial Apparatus

Much commercial apparatus is available for GLC. The range of prices and designs is large. Most of these consist of a packaged system made up of gas-flow controls, a sample inlet (with heaters), a column oven, a detector, and a recorder. Various types of columns and column packing are available as well as the range of detectors mentioned previously. The various companies listed in the Appendix are willing to supply additional information about their specific apparatus.

The remainder of this section will be devoted to a general discussion of the various parts of a GLC system and their operation. These topics will include: (1) the carrier gas and its flow rate; (2) sample injection of a solid, liquid, or gas with a syringe, and the amounts and precautions to be used; (3) the injection port, the best type of gasket, and how hot the port should be heated; (4) columns, what they consist of and their dimensions; (5) the solid support in a column and its requirements; (6) the stationary phase and its preliminary conditioning; (7) column construction; (8) the temperature of a determination, its choice and control; (9) the detector, a comparison of types and their sensitivities; (10) the attenuation of a signal; (11) recorders; and (12) two-column techniques.

Carrier Gas (The Moving Phase)

The specific carrier gases and their uses have been discussed in the previous section (p. 110), and are given in tabular form in Table 5.3 on p. 125. In this section we are more concerned with the mechanical control of the gases.

Some regulation of the gas pressure is necessary to keep a consistent gas flow. Fortunately, under most conditions, the reducing or diaphragm valve on a high-pressure tank can be used to give a sufficiently constant gas flow. Much of the newer GLC apparatus is equipped with sensitive needle valves designed for this specific operation, and the needle valves can be set to give an exact and reproducible flow rate.

The actual measurement of the gas flow can be done with a watch (stopwatch, if available) and a soap-bubble flowmeter. The bubble flowmeter is available commercially.[4] However, one can be constructed from a 10 ml graduated pipette, some soap solution, and a piece of rubber tubing. The end of the pipette is dipped into the soap solution, and this end is then pushed into the rubber tubing which is attached to the exit port of the GLC

[4] All GLC suppliers have them. See Appendix.

apparatus. The gas flow will push a soap bubble between the graduations on the pipette, and the length of time for a specific volume can readily be put on a ml/min basis.

The optimum amount of gas flow is dependent on the diameter of the column; a ¼ in. column usually requires 50–70 ml/min and a ⅛ in. column requires about 25–30 ml/min. The optimum conditions for any specific column can be, and at times should be, determined by relating gas flow to the resolution of two closely boiling liquids (for example, benzene and carbon tetrachloride), and plotting the results in some fashion. The curve should readily indicate the best gas flow. Figure 5.5 shows a curve obtained when the flow rate was decreased from 25 ml/min to 10 ml/min.

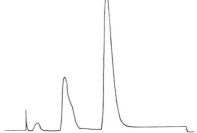

FIG. 5.5. A recording of a GLC chromatogram showing the increased separation and increased tailing when the gas flow was reduced from 25 ml/min in Fig. 5.3 to 10 ml/min.

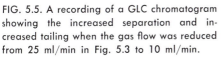

The original curve of the mixture being separated is shown in Figure 5.3.

Two major problems can arise when any gas flow is chosen; the flow can be either too fast or too slow. The shape of the peaks in the too-fast and too-slow situations can be poor, especially when high resolution and quantitative work is involved. In addition, if the gas pressure is too high or too low the hydrogen flame in the flame detector (if the apparatus is so equipped, see below) will either not ignite or not remain ignited.

Sample Injection

The basic objective in sample injection is to put the sample into the apparatus in as short a time as possible. It is reasonable to assume that a sample introduced over a ten-second period will give a smear on the column in comparison to a sample injected in one second or less. Figure 5.6 shows this broadening and loss of resolution of the peaks for *sec*-butyl alcohol and dioxane under improper injection conditions.

Although matter exists as solid, liquid, and gas, its introduction into the GLC apparatus can generally be reduced to two situations, liquid and gas. This is because a solid can either be dissolved in a low-boiling liquid that will pass through the column much faster than the high-boiling solid, or it can be dissolved in a liquid not normally detected by the detector (carbon

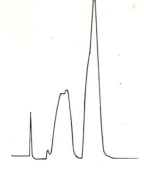

FIG. 5.6. A recording of a GLC chromatogram showing the effects of a slow sample injection. The chromatogram should be compared with Fig. 5.3.

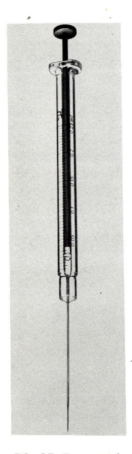 

FIG. 5.7. Two special syringes for the injection of GLC samples. The left-hand one is a 25 μl gas-tight syringe with a fixed needle and the right-hand one is a metal solids injection syringe. (Reproduced through the courtesy of the Hamilton Co., Inc.)

disulfide with a flame detector). However, special syringes are available [5] (Figure 5.7) which permit the direct introduction of a solid into the injection port. These syringes have both a metal barrel and a metal plunger; thus enough force can be generated with the plunger to force the solid onto the column.

The introduction of a gas can be carried out with a commercial gas sampling valve or a gas-tight syringe (Figure 5.7). The gas sampling valve attaches to the inlet port of the GLC apparatus and permits a measured amount of the gas to be injected as a "slug" into the column by using the pressure of the gas flow. For most qualitative work, however, a gas-tight syringe is adequate.

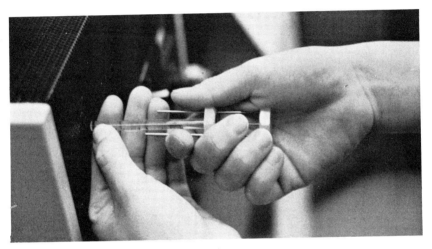

FIG. 5.8. Sample injection showing the use of a fixed-needle syringe equipped with a guide. Note that the cupped hand stops the plunger from being forced backwards when the needle enters the gas stream.

The syringe used for the gas sample can also be used for any liquid sample. Thus, it is desirable to have at least one gas-tight syringe for general GLC operation. The most often used type is the fixed-needle microsyringe, which comes in several designs and sizes. It is possible to use a removable-needle hypodermic syringe.[6] However, they usually deliver too much sample for the high-sensitivity (hydrogen-flame detectors, etc.) equipment. Thus, the purchase of a 10 μl and a 100 μl fixed-needle microsyringe is recommended, and the added expense of purchasing the 10 μl one with a Chaney adapter or guide (Figure 5.8) attached is justified.

[5] Hamilton Microsyringes, which can be obtained from any GLC supplier. See Appendix.
[6] Most commonly the Becton-Dickenson brand. See Appendix.

The amount of sample to be injected depends on the column and especially on the detector being used. Too much sample overloads the column, causes flooding, and decreases the resolution greatly. Frequently, placing a sufficiently small sample (10^{-6}–10^{-7} g) into the high-sensitivity equipment can be difficult. For this reason many liquids are injected as solutions and some apparatus (capillary columns) is equipped with a sample splitter where only a small amount (10^{-6} g) of the sample is permitted to enter the column. It is important to learn whether too much sample is being injected. This can be done by reducing the sample size and checking the resolution of the materials involved. To give some recommendations: the initial sample with a thermal conductance detector should be about 5–10 μl and that for the high-sensitivity detectors should be about 1 μl of a 1% solution.

Since many solutions involve low-boiling and sometimes moderately toxic and/or flammable solvents (carbon disulfide, both a toxic and a flammable solvent, is one of the best for use with a hydrogen-flame detector since the detector is insensitive to it), the proper handling of any sample becomes important. The use of bottles with serum caps or any type of rubber cap through which the needle of the syringe can be pushed to obtain a sample is recommended.

Injection Port

The injection port should be hot enough to vaporize the injected sample immediately. This is usually accomplished by having the temperature of the port 10–15° higher than the column. However, when the sample has one or more high-boiling materials, a higher temperature is often required. It is possible to cause a decomposition or rearrangement of the sample if the temperature is too high.

To test the efficiency of the injection system, it is suggested that the temperature be varied to determine whether better resolution results and whether the sample is being changed. If odd-shaped peaks result from a considerable increase in temperature, some breakdown in the sample may be occurring.

If only one heater is present for the injector port, column, and detector in the GLC apparatus, the temperature of the port is regulated by that of the column. Under these conditions the sample should be placed as close to the entrance of the column as possible. This practice is one that can be recommended for any GLC experiment.

Two additional suggestions can be made. First, silicone rubber septums should be used in the injection port. This rubber is one of the few that will withstand the high temperatures often used. Secondly, when a sample is injected, the plunger as well as the barrel of the syringe should be held. This stops the moderate pressures used in the gas train from forcing the plunger and sample out of the syringe when the sample is injected. Figure 5.8 shows the proper way to hold the syringe.

Columns

A GLC column, Figure 5.9, involves tubing filled with a solid coated with a liquid (to yield a **packed column**) or capillary tubing coated with a thin layer of liquid (a **Golay** or **open-tubular column**). In this discussion the solid will be consistently called the **support** and the liquid will be called the **stationary phase** or **liquid phase.**

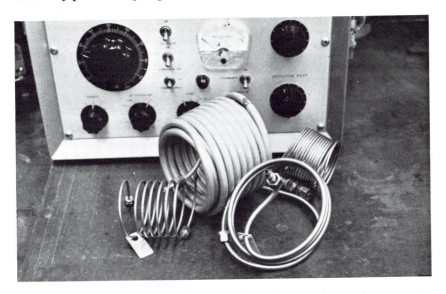

FIG. 5.9. Some representative GLC columns. The large-diameter column is for preparative work.

The tubing in a packed column can be stainless steel, aluminum, plastic, copper, or glass, all of which can be bent to a shape that fits into the oven in the apparatus. Stainless steel is most often used, for it is the most stable to heat and least reactive toward the samples being separated. However, for some separations, glass tubing is the only kind that can be used.

The internal diameters of the packed columns vary by a factor of about 400, for some preparative GLC columns are 4 in. in diameter while some analytical (Golay) columns are only 0.01 in. The column diameter does depend on the detector used in the apparatus, for a ¼ in. column is needed to separate the minimum amount of material (1–5% impurity) required for a definite peak with thermal conductivity detectors, and a ⅛ or ¹⁄₁₆ in. column is used with the more sensitive detectors. Since they are difficult to use and adverse mixing effects occur in the very large-diameter columns, a

½ in. column, Figure 5.9, and an automatic recycling system are often used for preparative GLC. See Section 5.6, p. 139 for more details.

[The length of a packed column is regulated by two factors. First, longer columns yield better separations. This is because more of the stationary phase will be in contact with the mixture of materials. Second, the longer the column, the greater the **pressure drop.** This pressure drop is the difference between the inlet and outlet gas pressures of the column and depends not only on length, but also the fineness (mesh) of the solid support. A large pressure drop normally makes optimum operation difficult.

Thus, the length of the column is a balance of the two main factors with the additional effect that the stationary phase may exert. These interact such that most packed columns are 4–12 ft long with a few as long as 30 ft.

Solid Support

The solid support is present to distribute the liquid or stationary phase evenly over a large surface area. The solid should be inert to avoid adsorption, stable to further crushing, uniform in size or mesh, and have a large surface area. Some form of diatomaceous earth is the most commonly used support material. Table 5.2 lists some of the diatomaceous and other supports currently being used.

TABLE 5.2

Solid Supports for the Stationary Phase in GLC Columns [a]

Diatomaceous	Nondiatomaceous
Chromosorb P, W, and G	Chromosorb T (Teflon)
Anakrom U, A, AB, ABS, AS, SD, P, and PA	TEE SIX (Teflon)
Anaprep U, A, and ABS	Fluoropak 80
C-22 Firebrick	Kel-F 330 LD
Celite 545	Glass beads
GC Super Support	Metal helices
Gas Chrom S and P	Tide detergent
	Poropaks (water)
	Molecular sieves (gases)

[a] The Chromosorb supports are trademarked products of the Johns-Manville Corp. and the Anakrom and Anaprep supports are trademarked products of Analabs, Inc.

When a support that contains a large number of reactive sites is used (a noninert solid), considerable tailing of the peaks occurs. To lessen or obliterate this adsorptive activity, the diatomaceous material (Chromosorb and Anakrom supports in Table 5.2) is treated with a deactivating substance. One such substance is hexamethyldisilane, which reacts with the active sites and puts a trimethylsilyl group on them to give a relatively inert support. This can then be used with the proper liquid coating. Most columns are now prepared with a deactivated support.

The Golay (capillary) or open-tubular columns can be made of the same materials as the packed columns, but stainless steel or glass tubing is most often used. Their inside diameter, since the capillary system is involved, is only 0.01–0.03 in. and their length is 50–500 ft. This length is necessary for the sample to come into contact with, and thus be resolved by, the proper amount of stationary phase. The amount of stationary phase is only 10–50 mg of liquid per 100 ft; thus the coating on the walls is very thin, 0.4–2 μ (10^{-6} in.). This length is possible because the pressure drop in these columns is very slight. The added problems connected with the use of this type of column, such as splitting the sample to use only about 10^{-6} g, warrants their use, for they rapidly separate complex mixtures and closely relat d isomers.

A recent development in capillary columns is one in which the stationary phase contains a small amount (5–10%) of finely divided (300–400-mesh) solid support. This, in effect, gives a much greater surface area of stationary phase with little change in the pressure drop and a net result of enhanced resolution.

Stationary Phase

The liquid or low-melting solid being used as the stationary phase must be nonvolatile under the conditions of the experiment. Thus, there will be a minimum temperature for a low-melting solid and a maximum temperature for every material. This maximum temperature commonly ranges from 100° to greater than 400°, depending on the material (see Table 5.1). Since every column will **bleed** some stationary phase when new, all columns should be **conditioned** before use. This is done by heating them 10–20° above the temperature at which they will be used with the carrier gas flowing for at least ten hours. The column should not be connected to the detector during the conditioning period, for any volatiles coming off could foul the detector. This same technique can be used for cleaning or reconditioning a column. If a column is to be used at a specific temperature much lower (or higher) than the temperature at which it was conditioned, a short reconditioning can be recommended.

The choice of a specific stationary phase was discussed in Section 5.2. Some examples are given in Table 5.1.

Column Construction

Assembling a column is a moderately simple operation after the coated support is available. The column is fabricated by plugging one end of a piece of tubing of the desired dimensions with glass wool and slowly filling it with the coated support. During the filling, the tube should be constantly tapped or vibrated to insure a firm, even packing. The filling end is then plugged and the tubing is bent to the proper shape. A column works best when it is straight. However, efficient oven design requires a box-like shape. A compromise results in columns in coils so that, at least, no sharp bends are present.

Amount of Stationary Phase.   Recent developments in preparing highly inert solid supports have permitted smaller amounts of the stationary phase to be used. Previously, enough stationary phase had to be coated on the solid support to cover all the active sites so that no adsorption could occur. Few columns now use the once common 30% stationary phase (by weight on the support), for most are now in the 5–10–20% range. The lower percentages will give a better resolution of the components of the sample when a small amount of sample is used. However, the small sample size is dependent on a highly sensitive detector. When a large amount of sample is required because of a low detector sensitivity, or for a preparative separation, the 20–30% liquid coating should be used to give adequate resolution.

If a commercial column is to be purchased for a specific need, it is usually best to obtain from the manufacturer recommendations for a suitable support, a stationary phase, the percentage of stationary phase, and the length of column.

Impregnation with Stationary Phase.   Several methods of coating the solid support are used. All involve dissolving a weighed amount of the stationary liquid phase (for example, 1.00 g of silicone oil) in a volatile solvent (any amount of hexane in this case), adding this solution to the solid support (10.0 g in this case), and allowing the solvent to evaporate. The result will be a support containing 10% stationary phase. This evaporation is done in three ways: by placing the slurry of solid support and solution in a rotating evaporator and removing the solvent with moderate vacuum (not with Chromosorb T and possibly others); by placing the slurry in an open tray or pan and allowing the solvent to evaporate; or by using a funnel coating technique. In this last technique the solid support is placed in filter paper in a funnel and the solution is poured over the solid. The solvent is allowed to pass slowly through the funnel and evaporate from the top of the support.

Glass and capillary columns sometimes require special equipment for their preparation, but it is possible to coat the simpler ones. The "plug" method involves introducing a specific amount of a 5–10% solution of the stationary phase in a volatile solvent into the column. This solution is slowly pushed through the column with an inert gas. During this operation the solvent evaporates leaving a film of stationary phase on the walls of the capillary. The "evaporation" method requires that the column be filled with the specific percentage of stationary phase solution, and that the solvent be slowly evaporated to leave a coating of liquid on the walls. The support-coated capillary columns, in which some solid material (extremely finely divided diatomaceous earth) is suspended in the liquid stationary phase, require special equipment for column preparation which is available only in commercial column manufacturing facilities.

Temperature: Choice and Control

The temperature at which a GLC separation is carried out is the most important variable to be considered, since it is easily manipulated and has a great effect on the separation. This is in contrast to TLC and column chromatography, which are essentially temperature-insensitive. The effects of temperature in GLC have been briefly discussed in Section 5.2.

The choice of oven temperature is influenced by the following facts, which are also illustrated in part by Figure 5.10.

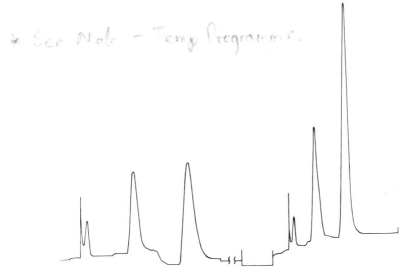

*See Note — Temp Programme.

FIG. 5.10. The effect of temperature on GLC separations. The curve on the left was obtained at 60° and the one on the right was obtained at 80° as in Fig. 5.3.

(1) Columns separate less effectively at higher temperatures. An increase of about 30° usually halves the solubility of a component of a mixture in the stationary phase, thus decreasing its retention time by half. Shorter retention times yield peaks that are closer together and poorly resolved.

(2) Columns operate more efficiently at higher temperatures. There is usually much less tailing and fewer poorly shaped peaks, mainly because of the increased volatilities of the components.

(3) The stationary phase could be eluted at too high a temperature. This bleeding will not only give spurious peaks, but will also destroy the column.

With these facts in mind, a temperature is usually chosen by experimentation which will give the required separation in a satisfactory amount of time. A good initial choice is one a few degrees lower than the boiling point of the major component or components of a mixture, if the average boiling point is known. If the history of the sample is so hazy that no boiling point is available, it is often best to obtain a boiling point. If all the materials have the same or nearly the same boiling point, the separation will be completely dependent on the nature of the stationary phase in the column.

The oven is controlled at a specific temperature by heat-input/heat-loss methods. Many commercial instruments and almost every "homemade" apparatus maintain the temperature by simple balancing methods. This is usually carried out by the trial-and-error adjustment of a variable voltage transformer which controls the heating coils. Sufficient voltage is applied to maintain the temperature at the desired point. This method is usually accurate to $\pm 2°$. More elaborate temperature designation and control can be obtained by the use of a direct dialing transformer. This controls an intermittent heater of a thermostatted oven and gives temperature regulation to better than $1°$. The latter type of equipment will normally be more expensive and is necessary only when the ultimate in resolution is required and expense is no object.

There is a further problem connected with temperature control when one oven contains both the column and a thermal conductivity detector. The problem occurs because the sensitivity of the detector changes with temperature. Under these conditions a changing recorder baseline associated with a change in temperature of the detector is a common occurrence and is alleviated by either controlling the temperature more accurately or attenuating the detector output to the recorder.

One final aspect of temperature control and change must be mentioned: the controlled change of temperature during a GLC run. This is an extremely powerful technique for enhancing the resolution on a column and can be used on mixtures with a wide boiling range and/or a large number of components. The technique and usefulness of temperature programming will be discussed in a separate subsection in Section 5.6 on special techniques, p. 142.

Detectors

The detector indicates the presence and measures the amounts of the components of a mixture in the effluent of a GLC column. Two major types of detectors are available, the integral and the differential, but since the latter gives the results more clearly, it is almost always used. The gas chromatograms pictured in these chapters are all from a differential detector.

Detectors of **moderate sensitivity** can be designed on a variety of principles, but few are used widely. The measurements on the effluent involve changes in gas flow, surface potential, density, dielectric constant, heat from burning, and thermal conductivity. The thermal conductivity method is the only one of these which is widely used, and will be the one described most completely.

The **high-sensitivity** detectors are based on electrical discharge, alpha- and beta-ray deflection and absorption, and flame ionization. The flame ionization detector is probably the most widely used and most versatile high-sensitivity detector and was used for the chromatograms in these chapters. These highly sensitive detectors permit a column to operate at high resolution, for only a small amount of sample is necessary for detection.

TABLE 5.3

Comparison of GLC Detectors

	Thermal Conductivity	Flame Ionization	Electron Capture
Minimum Quantity Detected	2–5 μg	10^{-5} μg	10^{-7} μg
Temperature Sensitivity	high	none	some
Carrier Gas	He	He or N_2	N_2 or Ar
Temperature Limit	450°	400°	225°
Response	all substances	all but H_2O and CS_2	not hydrocarbons, alcohols, ketones, acids
Special Use	for H_2O	water solubles	especially for halocarbons

Table 5.3 compares the three detectors most often used. The facts might suggest that one should obtain equipment which has a flame ionization detector. However, specific needs and uses require that other detectors be available. Some of these specific needs resulted in the development of the electron capture detector which is used for halogen-containing insecticides and the phosphorus detector for phosphorus-containing insecticides. The thermal conductivity detector is necessary for the measurement of the amount of water in a sample.

Thermal Conductivity Detectors. These detectors depend on the fact that heat is transported away from a hot body at a rate which depends on the composition of the gas surrounding the body. The heat flow or conductivity is thus dependent upon the rate of motion of the gas molecules which, in turn, is a function of their molecular weight. Therefore, the smaller the gas molecules, the higher their mobility and the faster the cooling. For this reason helium is generally used as a carrier gas, since it will give the maximum cooling.

The detector consists of a double arrangement of thermistors (good only to 150°), or platinum, or metal alloy wires (good to 400°), one in a pure carrier gas stream and the other in the column effluent stream. When the composition of the column effluent changes, the wire will heat in comparison to the standard. (For example, hexane conducts heat away from a hot wire about $\frac{1}{12}$ as well as helium). This heating of the wire will change the resistance of the wire, allow less current to flow, and cause an imbalance in a Wheatstone bridge. The current needed to balance the Wheatstone bridge is measured by the recorder. It is obvious that changes in the external temperature of the detector will influence its operation.

The simple apparatus pictured in Section 5.1, Figure 5.1, has a thermal conductivity detector made from a light bulb from which the glass bulb is removed. Other simple ones can be made from glow plugs and other devices of the same type which allow a current to flow and where the current flow can be influenced by the gas surrounding the electrodes.

The major precaution to take when operating a GLC apparatus equipped with this detector is to have the gas flowing when current is applied to the detector; see the basic instructions, p. 108. Different current amounts can be applied to the detector, from 10 to 300 mamps, but it has been found that the higher the current the more sensitive the detector. It is important to have a continuous current flow, even though a differential system is involved. When the above precautions are taken, the detector can be used for quantitative work, for the linearity of the response of the detector to varied concentrations is good.

Flame Ionization Detectors. The flame involved in these detectors is from hydrogen being burned to water. The hydrogen is from either a hydrogen tank or a commercial hydrogen generator, both producing the hydrogen at a steady rate. The hydrogen is mixed with air, which is introduced at a steady controlled rate, before the two gases reach the detector. When an organic compound is present the burning also produces carbon dioxide. These carbon dioxide molecules are ionized to give ions in the detector and the amounts of these ions are subsequently measured to give the signal, which can be highly amplified. High amplification is possible with this detector since the background noise level is low.

The operation of this detector is more difficult than that of the thermal conductivity type. The gas flow through the column, usually nitrogen at

15–35 ml/min, and the hydrogen-air flow through the detector (15–35 ml/min) must be balanced so that the column flow is about 1 ml/min greater than the detector flow. This will vary greatly, depending upon the geometry of the detector. The flame should be readily ignited with the ignitor button on the apparatus under these conditions. If the gas flows become unbalanced the flame either will not ignite, it will not stay lit, or there will be a considerable amount of noise. The last will be apparent as a jumpy baseline. Also, if the flame tip or chamber gets dirty from the bleeding of the column, the detector is difficult to clean. A noisy baseline may also result from the accumulation of column bleed in the flame tip or in the detector chamber.

The response of this detector to materials is linear; thus it is excellent for quantitative work. A further attribute of the detector is that it is insensitive to changes in external temperature, and it will not detect either water or carbon disulfide. Thus, standard solutions in carbon disulfide are often used for both qualitative and quantitative work. Extreme care must be exercised, for carbon disulfide is moderately toxic and very flammable. For these reasons it should always be used in a syringe bottle with a rubber cap.

Electron Capture Detectors.   In these detectors a signal loss rather than increase is measured, for the detector measures the number of electrons given off by a beta-ray source. The molecules of nitrogen or argon used as the carrier gas are ionized at a standard rate and they absorb electrons. This absorption is measured to produce the baseline. When organic materials are present they absorb electrons; thus a decrease in signal results.

Phosphorus Detector.   This detector is a flame ionization detector modified so that it will detect a minimum of 2×10^{-4} μg of a phosphorus-containing material. It is insensitive to other materials, thus giving a powerful tool for the specific analysis of phosphorus insecticides.

Attenuation

Every commercial GLC apparatus is equipped with a device called an **attenuator.** This is a variable resistance potentiometer of definite levels which permits the signal to be reduced to $\frac{1}{2}$, $\frac{1}{4}$, $\frac{1}{8}$, $\frac{1}{16}$, etc. of the original response, so that the output of the detector can be retained on the recorder chart. The linearity of the attenuator is usually good enough to permit it to be used for quantitative work. It is most often used when a small amount of impurity is being detected in a large sample, and an accurate ratio of the two or more materials is needed. To do this, the major component peak is attenuated to a small fraction and the impurity peak is unattenuated. A commercial attenuator is shown in Figure 5.11.

FIG. 5.11. The attenuator dial on a flame ionization detector. The numbers indicate what fraction of the signal is being transmitted by the detector circuit (4 on the dial means 1/4 of the signal, etc.).

Recorders

In a strip chart recorder, the signal produced by the detector of the GLC apparatus is amplified and recorded as a continuous trace on a piece of paper, according to its relative strength at any particular time. The simple apparatus in Section 5.1, Figure 5.1, was slightly different, for it did not utilize a recorder. It required the manual measurement of the strength of the signal at a particular time, followed by a manual plotting of the data. The latter type of recording does not readily reveal the balance of the GLC apparatus, which is normally indicated by the level baseline on the strip recorder.

It is evident that the automatic recorders are almost a necessity when many chromatograms are to be run. This has resulted in the manufacture of several specific types of recorders for GLC, one of which has automatic attenuation of the signal so that the output always stays on the chart. Others are equipped with an integrator which facilitates obtaining data for quantitative work.

The most widely used recorder is one on which full-scale deflection of the pen is equivalent to a signal of 1 mV. This has the high sensitivity needed with the thermal conductivity detectors. The attenuator on the

GLC apparatus can be used to give a fraction of the signal when the more sensitive detectors are used.

Two-Column Techniques

Previously in this discussion, we have been describing work with a single column containing a single stationary phase. A second and powerful technique can often be utilized: that of using two (or more) columns in series. This gives a partial separation of the components of a mixture in one column with the complete resolution in the second (or third) column. The technique requires that the oven of the apparatus be large enough to accommodate two columns. If this is not possible, one can connect two GLC systems with metal tubing. However, a considerable amount of mixing can occur if the connecting tube is too long. This can be compared with two-dimensional TLC.

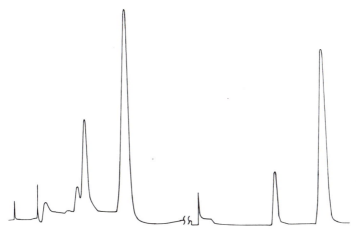

FIG. 5.12. Recordings of two chromatograms which show the effect of two columns placed in series for the separation of sec-butyl alcohol and dioxane. The curve on the left has a 5 ft Carbowax column followed by a 5 ft nonyl phthalate column. The curve on the right shows the results when the order of the columns is reversed.

The choice of the two (or more) columns depends on the constituents of the mixture involved. There are four possible arrangements of two columns: nonpolar, nonpolar; nonpolar, polar; polar, nonpolar; and polar, polar. The fact that there are stationary phases of intermediate nature, i.e., a phthalate column, gives many additional arrangements. The influence of the order of the two columns is shown in Figure 5.12, which indicates that one arrangement (on the left) gives enhanced separation.

This technique will often give separations not possible by single column methods and also permit separations which can be done on one column to be done more rapidly. This is especially true when the GLC apparatus is one that permits the two columns to be placed in different ovens, which can then be heated to either the same or different temperatures. The latter is important when the two stationary phases have different maximum temperatures. There is a way of obtaining the effect of two columns with one column. This one column is made with a mixture of the two stationary phases in the proportions present in the two columns. This arrangement does give capability, but with a loss in versatility.

5.4. THE IDENTIFICATION OF COMPONENTS IN A MIXTURE

"Pure" materials have often been shown to be impure by GLC; see Figure 5.3. A chromatogram from a high-resolution GLC apparatus can indicate the presence of one or many impurities in the parts per million range. In fact, chromatographic techniques have shown that many supposedly pure materials were quite impure. On occasion, these impurities have been found to influence the direction of a chemical reaction. The identification of every component shown by GLC is often impossible, mainly because of the time and difficulties involved and the minute amounts of material which can be detected.

This section will describe some chromatographic and nonchromatographic methods available for the identification of materials in a mixture. The chromatographic methods are: (1) retention times, (2) the addition of the supposed material, (3) logarithmic plotting, (4) the use of two columns, and (5) the use of two detectors. The nonchromatographic methods are those which result from a chemical reaction on the effluent or from a physical measurement on the isolated or **trapped** fraction.

Chromatographic Methods

Retention Times.   The retention time is the time, in minutes or seconds, elapsing between the time that a standard air peak emerges from a chromatographic column at a given temperature and gas flow and the time that a given substance emerges under the same conditions. In a sense, it is analogous to the R_f value used in TLC and it is usually about as reliable. For the identification of substances by GLC, it can serve only as a first approximation. This unreliability is due to the number of variables involved even when working with the same column. Two of these variables are the "age" of the column and the influence of the other mixture components on the retention time of the substance being measured. The major problem in repeating work from the literature is the exact duplication of column conditions.

Addition of Suspected Substance (*Spiking*).   This technique involves adding a small amount of a known material to a portion of the sample being investigated followed by an examination of the subsequent chromatogram. If the peak of interest is enhanced, the identification is positive (but not definite; see below). However, if a new peak or **shoulder** results, the identification is negative. Figure 5.13 illustrates the method.

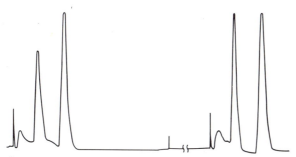

FIG. 5.13. Recordings of two gas chromatograms which illustrate the technique of spiking. The first is the sec-butyl alcohol–dioxane mixture shown first in Fig. 5.3. It was suspected that the third peak from the left was sec-butyl alcohol. This was confirmed by adding a portion of pure alcohol to the mixture and observing the enhanced peak in the second chromatogram.

This frequently used method depends upon the facts that a completely unknown sample is rarely examined by GLC and that a number of compounds are available for comparison. Often, smelling a sample will indicate which class or classes of organic compounds are involved, or some chemical test or physical measurement (infrared) can be used (see below). This knowledge and the knowledge of the boiling point necessary for choosing the column temperature will often give enough information to suggest what the major component(s) might be.

A GLC analysis will also give information about the impurities in the mixture, especially, depending on the column used, specific knowledge on whether they are lower or higher boiling or more or less polar than the major component(s). For example, if a nonpolar liquid phase such as an Apiezon grease is used, the peaks with a shorter retention time are usually from compounds which are lower boiling than the main component, and vice versa. The lower boiling points can result from their having fewer carbons or being more highly branched. When a highly polar liquid phase such as a Carbowax is used, the less polar components have lower retention times and vice versa.

If the first attempt at identification is positive, be cautious. It often happens that two compounds give the same behavior on a specific column at the conditions that are being used. Thus, for absolute identification, it is necessary to show the peak enhancement at a different temperature and flow rate, but better still, on a different column, preferably one of greatly different polarity. The use of a chromatographic and a nonchromatographic method is often the best.

Logarithmic Plots. An unknown member of a homologous series of compounds can be identified by the relationship of its retention time to those previously measured. Charts can be prepared which compare the logarithm of the retention time (actually retention volume, but this does not matter at constant gas flow) with the number of carbon atoms (or other factors) at constant conditions, to give an almost straight line plot. This can only be done within a homologous series of compounds; thus, normal alcohols, iso-acids, specific esters, etc., give different plots.

This technique requires that the known materials be available and that the conditions for a separation be reproducible. The latter aspect can be difficult to obtain, for the column and other variables do change with time. It is an especially good method when a large number of analyses of a specific group of materials must be carried out.

Two Columns. This was described above, but will be restated. To obtain an absolute identification, it is highly recommended that the technique of the addition of a suspected material be done on two different stationary phases of different polarities. If the particular separation involved two columns in series, p. 129, one or both of the columns should be changed.

Two Detectors. An unknown substance will give the same recorder response as the known material to two different types of detectors in the same environment if it is, in fact, identical to it. Almost any combination of detectors can be used; flame ionization and thermal conductivity; flame ionization and electron capture; thermal conductivity and "sniffing" the effluent, etc., but not all combinations are practical. The use of a moderately sensitive and a highly sensitive detector together can give problems since they require different sample sizes.

To facilitate the identification of materials by this method, GLC equipment is available which has dual channels from the column to the detectors. The outputs from the two detectors can be recorded on two one-channel recorders or one two-channel recorder. This permits a rapid comparison of the responses to a particular material.

Nonchromatographic Identification

Functional Group Classification Tests. These tests require that the effluent constituting a specific peak be bubbled into a solvent containing a specific reactant.[7] This can be done by splitting the effluent into many

---

[7] J. T. Walsh and C. Merritt, Jr., *Anal. Chem.* **32,** 1378 (1960).

streams. A simple apparatus for this type of operation can be made by forcing syringe needles through some type of rubber bulb, which in turn can be attached to the output port. The syringe needles can be spaced such that they can readily be placed in small test tubes containing the different reagents. The reagents are the normal qualitative organic reagents found in any text on the subject. Alternately, the effluent peak can be passed through just one solution at a time. The latter would require that a number of GLC separations be carried out (one for each solution).

The reagents are those that give specific chemical tests, usually by a color change, for an alcohol, aldehyde, ketone, amine, unsaturated hydrocarbon, etc., and do serve to indicate the class of compound or the undetected presence of another class of compounds in a major peak. This type of test can be run on a number of fractions of the effluent gas, and it is generally used in conjunction with more specific identification techniques.

The functional group test is especially useful when little information is available about a mixture and about the particular impurities in a mixture. The amounts necessary for most of the reactions (20–100 μg) require that a high-capacity thermal conductivity detector be used or an apparatus with a highly sensitive detector and a stream splitter. For example, the stream splitter would be needed with a flame ionization detector, for the sample is completely destroyed in this detector.

Sample Pretreatment.   Complex mixtures are frequently chemically pretreated before chromatography with specific organic reagents. For example, extraction of the sample with sodium bisulfite will remove aldehydes and methyl ketones; aldehydes and ketones can be reduced with sodium borohydride;[8] or treatment with calcium hypochlorite will remove methyl ketones (by the haloform reaction). If the chromatogram of the treated sample is compared with the chromatogram of the untreated sample, the functional groups present in the various peaks of the original mixture can frequently be surmised.

Derivative Formation.   This involves the reaction of a specific peak in the effluent with a specific type of reagent which forms a crystalline solid upon reaction. The functional group of the material must be known (see above method) to choose the correct reagent and a sufficient amount of solid must be obtained to be able to determine the melting point. The melting points of many materials are listed in specific books and would permit an identification of the sample.

Trapping of the Fraction.   The effluent from the GLC column is in the vapor phase, and if this is condensed in a receiver by cooling, the unknown can be trapped.

This trapping is very often done by putting a thin-walled glass capillary tube into the exit port of the column at the correct time and permitting the

[8] M. E. Morgan and R. L. Pereira, *J. Dairy Sci.* **46**, 1420 (1963).

cooler walls of the capillary to condense the liquid (or solid). Again, this requires a thermal conductivity detector or a stream splitter with a highly sensitive detector to obtain enough sample.

Another method of trapping involves allowing the effluent to pass into a cooled test tube, or better, a centrifuge cone. The amount of cooling can be regulated by using cold water, ice, Dry Ice, or liquid nitrogen. There are a number of commercial devices that can be obtained which efficiently carry out the trapping of a specific peak or many peaks.

The trapped sample can be examined by a number of powerful instrumental techniques. The main ones are: infrared spectrometry, nuclear magnetic resonance, and mass spectrometry. Many others could also be listed. Infrared is very useful because some 30,000 spectra of known compounds are presently available for comparison (through the Sadtler Spectra).

Mass spectrometry can also be carried out without an actual isolation of the sample. Many GLC systems have been constructed which permit each fraction or peak to pass directly into the mass spectrometer. The data resulting from this measurement along with that obtained from the addition of the suspected substance are sufficient for most identifications.

5.5. QUANTITATIVE ANALYSIS OF A MIXTURE

One of the major features of GLC, and this was mentioned in Chapter 1, is that it not only indicates the number of components in a sample, but it also gives some idea of the amount of each component. The first estimate of amount may come from a visual comparison of the areas under the peaks in a chromatogram recording. Under proper conditions, the careful integration of the areas will give data with an accuracy of better than 1%.

The operation of a GLC apparatus for quantitative work requires that the temperature, gas pressure, and other operational features remain more constant than that needed for qualitative work. Basic details on carrying out these operations have been given in the previous chapters; this chapter will consider the special requirements.

The additional data required to obtain accurate quantitative results are (1) the areas of the peaks and (2) an accurate correlation of the areas with the amounts of the components. This chapter will describe the methods used to measure or estimate the areas and to determine the *sensitivity* or *correction* factors which relate the areas to the amounts.

Area Measurement

Area measurement can be done by four general methods: [9] (1) by cutting and weighing, (2) by triangulation, (3) by using a planimeter, and

[9] See Varian Aerograph's Review of 7/67 on this subject.

(4) by direct integration. When precision and speed are considered, some type of direct integration of the peaks is best.

Cutting and Weighing.   This method involves cutting out the peaks from the recorder sheet with scissors and weighing the paper shapes on an analytical balance. The weight of a specific peak as compared to the weight of all the peaks gives the weight percentage of the fraction or fractions.

This is a somewhat tedious procedure and is less precise (one-half to one-fifth) than instrumental integration. The major errors that give the poor precision arise from poor cutting techniques and variations in the thickness and moisture content of the paper, especially when small peaks are involved. In addition, the record of the chromatogram is destroyed. The use of a Xerox copy (or similar photocopy) of the recorder sheet (where the Xerox paper has greater homogeneity) affords the precision mentioned above and also allows the retention of the record.

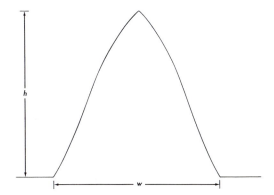

FIG. 5.14. A determination of the area of a symmetrical peak resulting from a gas chromatogram, $a = hw/2$.

Triangulation.   This method assumes that the peak is a perfect triangle (Gaussian or symmetrical peak) and obtains the area by the formula: $a = hw/2$, Figure 5.14. Obviously, when a peak is unsymmetrical (tailing or leading) the formula will not hold. The precision is about one-half to one-fifth that of an integrator.

Since all of the peaks in a spectrum are rarely symmetrical, it is necessary to be able to estimate the area of an unsymmetrical peak. This is done by using the formula: $a = (w_1 + w_2)h/2$, Figure 5.15.

The triangulation method is often used when two peaks overlap, or in trace analysis. Rarely will the analysis with overlapping peaks be highly accurate, but when both peaks are symmetrical they can be evaluated with the formula above and a series of known mixtures. Trace analysis is most

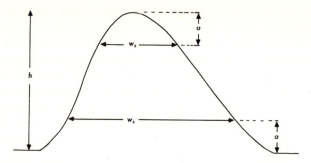

FIG. 5.15. A determination of the area of an unsymmetrical peak resulting from a gas chromatogram, $a = (w_1 + w_2)h/2$, where w_1 is the width of the triangle not at the base, but at the point where the curve has a flex point, and w_2 is the width of the triangle measured at a point the same distance (a) from the top of the peak as w_1 is up from the base.

FIG. 5.16. Trace analysis situations. The chromatogram recording on the left shows that an impurity which precedes the major component out of the column can be accurately measured. The chromatogram recording on the right shows the impurity following the major component and the difficulty which arises.

accurate when the trace material precedes the major component; for when it follows, the major component usually tails enough to make an accurate measurement difficult (see Figure 5.16).

Planimeter.   A planimeter is an instrument for measuring the area of a plane figure by tracing its boundary line. The original cost for a good planimeter is about one-third that of an integrator. However, its use is tedious, often slow, and requires considerable practice. Its precision is often insufficient (one-fifth to one-tenth that of an integrator) for measuring small amounts of material.[10]

Integrator.   The integrator is a device attached to the recorder which indicates the area of a peak by a measurement based on the displacement of the slidewire in the recorder (mechanical) or on the current generated by the detection system (electronic). Both give a high degree of accuracy.

[10] A planimeter is available from companies selling draftsmen's supplies.

The latter is not only easier to use, but also gives about twice the precision of the former.

The major mechanical integrator in use is the Disc Integrator,[11] which is attached to the recorder and gives a trace on the chart paper opposite each peak (see Figure 5.17). This trace has been on all of the chromatograms pictured in the chapter on GLC, but was not included in the drawings. This mechanical integrator is designed so that the pen speed is proportional to the displacement of the recorder pen from the baseline; thus

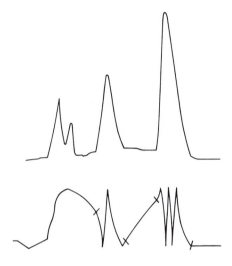

FIG. 5.17. A recording of a gas chromatogram of the 1:2 mixture of sec-butyl alcohol and dioxane. The integrator trace below the peaks has been enhanced to facilitate evaluation. Appropriate starting and stopping points on the integrator trace are marked.

the number of counts (each complete trace between the limits is 100 counts) traced during the peak is proportional to the area of the peak. The areas of a number of peaks are compared by adding their total counts and calculating what percentage of the total is from each peak. In Figure 5.17 the count for the alcohol peak is 247 and for the dioxane 516. This suggests that the dioxane is present in 68%, where its actual amount is 67%; however, see p. 139 for more details.

The major problem in using the Disc Integrator for quantitative GLC is that the operation must be closely attended. It is necessary to attenuate (p. 127) any peak that goes off scale to determine the multiplier for the

---

[11] Disc Instruments, Inc.

number of counts for that specific peak. For example, if only one peak in a chromatogram was attenuated, and it to one-fourth of its signal, the number of recorded counts for that peak would have to be multiplied by four. However, since some error in counts can arise during attenuation, two chromatograms must be carried out to obtain accurate data. One will show the necessary attenuation factor for each peak and the second will allow attenuation to be changed to the proper factor at the proper time, which is when the peak is to begin. Multiplying the number of counts by the attenuation factor will give the most exact area measurement.

The use of the considerably more expensive (about ten times the mechanical) electronic integrator is currently the most accurate (two to three times the mechanical) and easiest method of quantifying GLC data.[12] There is no necessity for attenuation, since the strength of the signal is read directly by the electronic device. A direct number readout for each peak and for the total of all the peaks is given, and the percentages can be obtained directly.

Related to the electronic integrator is the direct computer recording and printout of the data. The data can also be recorded on punched cards, on paper tape, and on magnetic recording devices. This type of operation is prohibitively expensive for most laboratories, but is being increasingly used by the major oil and chemical companies.

Calibration Procedures and Correction Factors

It might be expected that the areas of the peaks in a chromatogram are directly related to the amounts of the substances. However, this is rarely exactly true in practice. Therefore, in every determination it is necessary to calibrate the system by relating the amount of material to the area of the peak produced. Two main methods for this calibration are available: one involving correction factors and one involving blends of the materials.

Correction Factors. The sensitivity or correction factor compensates for the difference of response of the detector to the type or class of compound. It is expected and well known that the response of the thermal conductivity detector to an ether will be different from that to a cyclic ketone. However, in any case one would expect the response to be linear for a specific compound.

The correction factors can be large, i.e., changes of 100% are found, but they do give a method of rapidly and consistently correcting an area to the amount of component. These correction factors can also be based upon the use of one component in the mixture as a standard and relating the other factors to it.

The correction factors are measured by preparing a solution of known

[12] Most of the companies selling GLC equipment have one for sale. See Appendix.

concentrations of all of the compounds present in the unknown, obtaining its chromatogram, and relating the areas to the known concentrations. It is best if the components in the known are in similar ratios as the unknown, and it is imperative that the GLC conditions be the same. The ratios within the sample must be similar because one material will influence the response of another and this influence is greater when the ratios are quite different. For this reason a correction factor of a compound in one mixture is not applicable to another mixture.

The integration data on Figure 5.17 will be used to illustrate the use of correction factors. As it was indicated on p. 137, the number of counts for the *sec*-butyl alcohol is 247, that for dioxane is 516, and the fractions were 0.32 and 0.68 respectively. Since 0.33 parts (by weight) of *sec*-butyl alcohol had been used to prepare the mixture, the weight correction factor is 0.32/0.33 or 0.97 and that for the dioxane is 0.68/0.67 or 1.02. This indicates that in future determinations of this mixture the detector will record only 97% of the amount of alcohol and 102% of the amount of dioxane, and that any quantitative weight data must be corrected with these predetermined factors.

Blends.   This time-consuming method uses a series of blends of the materials in the mixture which are chromatographed under the same conditions as the unknown. In fact, the unknown is often measured in the middle of the chromatograms from the blends. The area for each material in the various blends is determined and plotted against the amounts, and these calibration curves are used to determine the amounts of each compound in the unknown.

5.6. SPECIAL TECHNIQUES

The previous sections described many general features of GLC. However, there are some procedures and topics which should be described separately. These special topics are preparative GLC, pyrolytic GLC, temperature-programmed GLC, and methods for speeding up GLC. These will be described in order.

Preparative GLC

This technique permits the high separation efficiency of GLC to be applied to the problem of fractionating large amounts of closely boiling materials (less than 1° boiling range) and materials that normally form azeotropes (constant boiling mixtures of fixed proportions). Three possible ways are available for increasing the capacity of GLC equipment: the use of a single column with a large bore, the utilization of a cycling operation with a slightly increased column diameter, and the use of parallel columns of a slightly larger bore.

Enlarged Diameter Column.   This is logically the first step in increasing column capacity, for a larger amount of stationary phase will permit more material to be separated. Columns as large as 4 in. in diameter have been used, but a better choice is the 0.5–1.0 in. range. This will permit the separation of samples of 1 g or more.

Most standard GLC apparatus equipped with a thermal conductivity detector can be used for a preparative separation when fitted with a 0.5 in. column rather than the usual 0.25 in. column. The separation will require

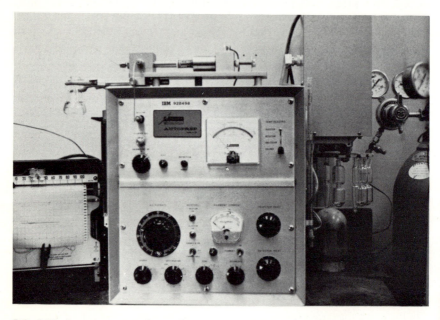

FIG. 5.18. An automatic cycling collector for preparative GLC. The device in the right of the picture rotates to collect fractions in the glass tubes. It is connected to the detector circuit so that it will rotate between peaks. The instrument is a Varian-Aerograph apparatus.

a greater volume of gas to maintain the 60 ml/min gas flow. Thus, the high-pressure gas cylinder should be checked before starting to make certain that a sufficient amount of gas is available. On occasion, the gas flow must be increased to 90 ml/min to give a good separation in a reasonable amount of time. The sample, usually about 1 ml, is injected in the normal manner, and the fractions are trapped as described above on p. 133.

Automatic Cycling Apparatus.   This type of equipment is pictured in Figure 5.18 and is highly versatile. A 0.5 in. column is normally used with a moderate flow rate. Samples up to 2 ml are injected. The recorder is coordinated with a rotating collection device such that after a peak has been recorded and the sample collected, the collector will switch to a clean vial for the next peak. After the first sample has been separated and collected, another sample is injected into the column automatically and separated into the same vials as the first sample and so on. An unlimited amount of sample can be separated by this method.

The particular apparatus pictured has a thermal conductivity detector, but equipment is also available with one of the high-sensitivity detectors and a stream splitter. Both types of systems are equipped with an automatic temperature programming and cooling cycle and the entire unit is designed for unattended operation.

The major problem with this type of equipment is that some part of the cycle could fail when unattended and this failure would not be detected until the operation was complete. A failure would require that the original separation be redone or that a particular fraction be rechromatographed.

Parallel Columns. Instead of having one column of approximately 2 in. in diameter, this type of apparatus uses eight (or more, or less) 0.5 in. columns in parallel. The flow rates must be adjusted so that the effluent from each column reaches the common exit port at the same time. This apparatus has the added advantage that the columns can be connected in series as well as parallel; thus four columns of twice the length, etc. can be used to separate a particular mixture.

Pyrolytic GLC

High-melting/high-boiling compounds and polymers cannot be examined by regular GLC. However, if these substances are broken down or decomposed to smaller fragments which are volatile, they can be identified on the basis of these fragments. This controlled decomposition of solids or liquids and the chromatograms produced are the basis of pyrolytic GLC.

The material to be pyrolyzed is placed either in a platinum boat or on a platinum coil in a commercially available attachment on the injection port of the regular GLC equipment, usually one with a flame ionization detector. The sample is heated in the gas flow to the proper temperature (500–900°) in a short amount of time, either by passing a current through the platinum coil or by some other rapid method of heating. The moving phase forces the pyrolysate on the column which gives a chromatogram in the normal way. The amount of heating causes different types and extents of decomposition. Thus, the temperature choice is important and must be consistent.

It is possible to relate the fragments produced from polymers to the monomeric units which make up the polymer, making a determination of the structure possible. The monomeric materials will give the same retention time when injected in the normal manner.

Temperature-Programmed GLC

It is often difficult, if not impossible, to obtain a complete separation of components when the boiling point *range* of a mixture exceeds 50° and also when many components are present. To alleviate this problem the technique of temperature programming was developed. This involves heating the column from one temperature to another at a controlled rate, as compared to the constant temperature used in regular GLC operations.[13]

The controlled change of the rate of heating can be done in a number of ways. It can be (1) linear at any chosen rate, (2) stepwise, (3) multilinear, i.e., several rates, (4) linear-isothermal, and (5) isothermal-linear. Curves illustrating each of these are given in Figure 5.19. The most fre-

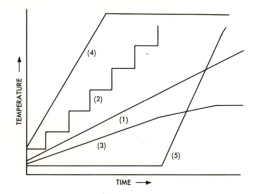

FIG. 5.19. Five types of temperature programming: (1) linear, (2) stepwise, (3) multilinear, (4) linear-isothermal, and (5) isothermal-linear.

quently used method is linear programming, with the two mixes of linear programming and isothermal operation next in use. The range of the temperature change is most often 50–150°, but temperatures up to 250° are used.

The major instrumental problem in programmed-temperature GLC is the effect of temperature change on the detector. This is not a problem with a flame ionization detector, but a thermal conductivity detector must be not only in a separate oven but also in a highly insulated one. The choice of initial and final temperatures is dependent only on the lower and upper temperature limit of the column or columns involved and the sample boiling point range.

A GLC apparatus that is not equipped with temperature programming

[13] W. E. Harris and H. W. Habgood, "Programmed Temperature Gas Chromatography," John Wiley & Sons, Inc., New York, 1966.

can be equipped by purchasing separate temperature programmers. The programming can also be done in a flame ionization system by setting the variable transformer which heats the column at some voltage and turning on the heater at a certain time. The increase in temperature of the column will occur at a surprisingly consistent and reproducible rate.

Speeding Up a GLC Determination

The following ways of speeding up a GLC separation are listed in a random order. Many are used together, so that a number of small changes can greatly increase the speed of a determination.

(1) Higher column temperature. An increase in temperature of $30°$ doubles the rate of throughput (but may decrease resolution).

(2) Decrease in amount of stationary phase. The decrease in amount is not linear with throughput, but is related.

(3) Increased gas flow. The gas flow may be less than the optimum amount.

(4) Programmed-temperature techniques. These are often used not only for speeding up a chromatogram, but also for getting a better separation.

(5) Dual column techniques. These permit a higher temperature and flow rate to be used to obtain the same separation.

(6) Capillary or open-tubular columns. These will often give a separation in seconds for which a packed column requires minutes.

6

The Literature
of Chromatography

INTRODUCTION

A large number of articles and books have been published on the general subject of chromatography and on the specific techniques which have been discussed in this book. These texts extend our discussions into practical and theoretical areas of considerable complexity. It would be quite reasonable to state that with the background provided by these additional sources, it should be possible to separate any mixture of any type.[1]

There are also several other sources of useful information which can be easily consulted. The Reviews of Analytical Chemistry is published each April by the journal *Analytical Chemistry*. Various topics are reviewed each two years. For example, gas chromatography and chromatography in general were reviewed in 1964 and 1966. The *Journal of Chromatography* not only publishes papers on the various aspects of chromatography, but publishes a running bibliography of papers that is divided into the various techniques and even into classes of compounds within the techniques. There is, in addition, a *Journal of Gas Chromatography* and a journal entitled *Separation Science,* which publishes more fundamental papers on chromatography. The volumes of "Advances in Chromatography" edited by J. C. Giddings and R. A. Keller present a series of review chapters on various aspects of the subject.

A further source of information lies in the bibliographies and literature which have been compiled and published by the various manufacturers of

---

[1] One possible exception to this statement would be the separation of a mixture of enantiomers. However, on properly impregnated supports, even this type of mixture has been partially separated.

144

chromatographic apparatus and materials. These are freely available to anyone interested in the technique.

CHROMATOGRAPHY IN GENERAL—REFERENCE BOOKS

1. E. Heftmann (ed.), "Chromatography," 2nd Ed., Reinhold, New York, 1967.
2. H. G. Cassidy, "Fundamentals of Chromatography," Interscience, New York, 1957.
3. H. H. Strain, "Chromatographic Adsorption Analysis," Interscience, New York, 1942.
4. E. Lederer and M. Lederer, "Chromatography, a Review of Principles and Applications," 2nd Ed., American Elsevier, New York, 1957.
5. T. I. Williams, "The Elements of Chromatography," Philosophical Library, New York, 1953.
6. A. T. James and L. J. Morris (eds.), "New Biochemical Separations," Van Nostrand, London, New York, 1964.
7. D. Abbott and R. S. Andrews, "An Introduction to Chromatography," Longmans, London, 1965.
8. R. Stock and C. B. F. Rice, "Chromatographic Methods," 2nd Ed., Chapman & Hall, London, and Barnes and Noble, Inc., New York, 1963.

THIN-LAYER CHROMATOGRAPHY—REFERENCE BOOKS

1. J. G. Kirchner, "Thin-Layer Chromatography," Interscience, New York, 1967.
2. E. Stahl (ed.), "Thin-Layer Chromatography," 2nd German Ed., Springer-Verlag, Berlin, 1967; 1st English Ed., Academic Press, New York, 1965.
3. K. Randerath, "Thin-Layer Chromatography," 2nd German Ed., Verlag Chemie G.m.b.H., Weinheim, 1965; 2nd English Ed., Academic Press, New York, 1966.
4. J. M. Bobbitt, "Thin-Layer Chromatography," Reinhold, New York, 1963.
5. E. V. Truter, "Thin Film Chromatography," Cleaver-Hume Press, London, and Interscience, New York, 1963.
6. Y. Hashimoto, "Thin-Layer Chromatography," Hirokawa Publishing Co., Tokyo, 1962.
7. M. Ishikawa, S. Hara, T. Furuya, and Y. Nakazawa, "Thin-Layer Chromatography, Fundamentals and Applications," Nanzando Co., Ltd., Tokyo, 1965.

8. L. Labler and V. L. Schwarz, "Thin-Layer Chromatography," American Elsevier, New York, and Publishing House of the Czechoslovak Academy of Sciences, Prague, 1965.
9. G. B. Marini-Bettòlo (ed.), "Thin-Layer Chromatography" (symposium 1963), American Elsevier, New York, 1964.
10. I. Smith and J. G. Feinberg, "Paper and Thin-Layer Chromatography and Electrophoresis" (A Teaching Level Manual), 2nd Ed., Shandon Scientific Co., Ltd., London, 1965.
11. "Thin-Layer Chromatography" (a collection of reprints of articles in *Lab. Practice*), United Trade Press Ltd., London, 1964.

THIN-LAYER CHROMATOGRAPHY—ARTICLES

1. R. Maier and H. K. Mangold in Vol. 3, "Advances in Analytical Chemistry and Instrumentation," C. N. Reilley (ed.), Interscience, New York, 1964, p. 369.
2. H. K. Mangold, H. H. O. Schmid, and E. Stahl in Vol. 12, "Methods of Biochemical Analysis," D. Glick (ed.), Interscience, New York, 1964, p. 393.
3. H. K. Mangold, *J. Am. Oil Chemists' Soc.* **38,** 708 (1961); **41,** 762 (1964).
4. E. Demole, *Chromatog. Rev.* **1,** 24 (1958); **4,** 26 (1962).
5. E. G. Wollish, M. Schmall and M. Hawrylyshyn, *Anal. Chem.* **33,** 1138 (1961).
6. E. G. Wollish, *Biochem. J. Sympos.* **2,** 687 (1962).
7. D. C. Malins and J. C. Wekell, *J. Chem. Ed.* **40,** 531 (1963).
8. L. F. Fieser, "Organic Experiments," D. C. Heath & Co., Boston, 1964, pp. 288–297.

COLUMN CHROMATOGRAPHY—BOOKS AND ARTICLES

There do not appear to be any books on the specific subject of column chromatography in the same sense that there are on TLC and GLC. However, the following sources can be consulted in addition to those cited in the category "Chromatography in General" given above.

1. R. Neher, "Steroid Chromatography," American Elsevier, New York, 1964.
2. I. E. Bush, "The Chromatography of Steroids," Pergamon Press, London, 1961.
3. C. L. Mantell, "Adsorption," McGraw-Hill, Inc., New York, 1945.
4. "The Merck Index," 7th Ed., Merck and Co., Rahway, N.J., 1960, pp. 1575–1600 (for commercial adsorbents).

5. E. Lederer and M. Lederer, in Vol. 4, "Comprehensive Biochemistry," M. Florkin and E. H. Stotz (eds.), American Elsevier, New York, 1962, pp. 32–268.
6. E. Heftmann, *Anal. Chem.* **36**, 14R (1964); **38**, 31R (1966).
7. V. E. Tyler, Jr. and A. E. Schwarting, "Experimental Pharmacognosy," 3rd Ed., Burgess Publ. Co., Minneapolis, Minn., 1962, pp. 69–79.
8. N. D. Cheronis, J. B. Entrikin, and E. M. Hodnett, "Semimicro Qualitative Organic Analysis," 3rd Ed., Interscience, New York, 1965, pp. 131–170.
9. K. B. Wiberg, "Laboratory Technique in Organic Chemistry," McGraw-Hill, Inc., New York, 1960, pp. 149–178.

GAS CHROMATOGRAPHY—BOOKS

1. A. I. M. Keulemans, "Gas Chromatography," 2nd Ed., Reinhold, New York, 1959.
2. S. Dal Nogare and R. S. Juvet, Jr., "Gas-Liquid Chromatography," Interscience, New York, 1962.
3. W. E. Harris and H. W. Habgood, "Programmed Temperature Gas Chromatography," Wiley, New York, 1966.
4. H. M. McNair and E. J. Bonelli, "Basic Gas Chromatography," Varian Aerograph, Walnut Creek, California, 1966.
5. H. Purnell, "Gas Chromatography," Wiley, New York, 1962.
6. C. Phillips, "Gas Chromatography," Academic Press, New York, 1956.
7. H. P. Burchfield and E. E. Storrs, "Biochemical Applications of Gas Chromatography," Academic Press, New York, 1962.

GAS CHROMATOGRAPHY—REVIEW ARTICLES

1. B. A. Rose, *Analyst* **84**, 574 (1959).
2. C. J. Hardy and F. H. Pollard, *J. Chromatog.* **2**, 1 (1959).
3. S. Dal Nogare and L. W. Safranski, *Organic Analysis* **4**, 91 (1960).
4. R. L. Pecsok, *J. Chem. Ed.* **38**, 212 (1961).
5. L. S. Ettre and W. Averill, *Microchem. J.* **2**, 715 (1962).
6. A. V. Signeur, "Guide to Gas Chromatography Literature," Plenum Press, New York, 1964.
7. S. Dal Nogare, *Anal. Chem.* **38**, 61R (1966).
8. R. S. Juvet, Jr. and S. Dal Nogare, *ibid.,* **36**, 36R (1964).

GAS CHROMATOGRAPHY—REFERENCES TO SIMPLIFIED APPARATUS

1. C. L. Stong, The Amateur Scientist, *Scientific American,* June 1966, p. 124.
2. P. J. Cowan and J. M. Sigihara, *J. Chem. Ed.* **36**, 246 (1959).

3. H. Ven Horst, *ibid.*, **37,** 593 (1960).
4. F. Sicilio, H. Bull, III, R. C. Palmer and J. A. Knight, *ibid.*, **38,** 506 (1961).
5. C. W. Schimelpfenig, *ibid.*, **39,** 310 (1962).
6. J. McLean and P. L. Pauson, *ibid.*, **40,** 539 (1963).
7. R. E. Herbener, *ibid.*, **41,** 162 (1964).
8. S. Lowell and H. Malamud, *ibid.*, **43,** 660 (1966).

Appendix

The following list contains the majority of the suppliers and manufacturers of chromatographic equipment and supplies for the United States. A more complete and up-to-date listing can be found in the annual "Laboratory Guide," published by the American Chemical Society through the journal *Analytical Chemistry*.

Thin-Layer Chromatography—General Equipment

1. Ace Glass Inc., P. O. Box 688, Vineland, N.J. 08360.
2. Applied Science Labs., 135 No. Gill St., State College, Pa. 16801.
3. Brinkmann Instruments Inc., Cantiague Rd., Westbury, N.Y. 11590. (U.S. representatives of C. Desaga, E. Merck, and Macherey Nagel as well as manufacturers.)
4. Camag, A. G., Homburgstr. 24, Muttenz, Switzerland. (See Gelman and Thomas.)
5. Colab, Inc., P. O. Box 66, 1526 Halsted St., Chicago Heights, Ill. 60411.
6. C. Desaga, Hauptstr. 60, Heidelberg, Germany. (See Brinkmann.)
7. Gallard-Schlesinger Chem. Mfg. Corp., 584 Mineola Ave., Carle Place, N.Y. 11514.
8. Gelman Instrument Co., P. O. Box 1448, Ann Arbor, Mich. 48106. (U.S. representatives of Camag as well as manufacturers.)
9. Kensington Scientific Corp., 1165 67th St., Oakland, Cal. 94608.
10. Macherey, Nagel and Co., Düven, Rhld., Germany. (See Brinkmann.)
11. Mallinckrodt Chemical Works, Second and Mallinckrodt Sts., St. Louis, Mo. 63160.
12. E. Merck, A. G., Darmstadt, Germany. (See Brinkmann.)
13. MISCO (Microchemical Specialties Co.), 1825 Eastshore Highway, Berkeley, Cal. 94710.

14. Ohaus Scale Corp., 1036 Commerce Ave., Union, N.J. 07083.
15. Quickfit, Inc., 1 Bridewell Place, Clifton, N.J. 07014.
16. Carl Schleicher and Schuell Co., 543 Washington St., Keene, N.H. 03431.
17. Shandon Scientific Co., 515 Broad St., Sewickley, Pa. 15143.
18. A. H. Thomas Co., P. O. Box 779, Vine St. at Third, Philadelphia, Pa. 19105.
19. Warner-Chilcott, Instruments Div., 200 S. Garrard Blvd., Richmond, Cal. 94804.

Thin-Layer Chromatography—Specific Items

1. Aluminum Co. of America, 1501 Alcoa Bldg., Pittsburgh, Pa. 15219. (alumina)
2. Alupharm Chemicals, 610–612 Commercial Pl., P. O. Box 30628, New Orleans, La. 70130. (U.S. representatives for M. Woelm.) (adsorbents for TLC and columns)
3. Analtech, Inc., 100 S. Justison St., Wilmington, Del. 19801. (prepared layers)
4. Baird Atomic, Inc., 33 University Rd., Cambridge, Mass. 02138. (counting equipment)
5. Bio-Rad Laboratories, 32nd and Griffin Aves., Richmond, Cal. 94804. (adsorbents)
6. Calbiochem, 3625 Medford St., Los Angeles, Cal. 90063. (adsorbents)
7. Corning Glass Works, Technical Products Div., 3910 Crystal St., Corning, N.Y. 14830. (ground porous glass as a TLC adsorbent)
8. Dow Corning Corp., Midland, Mich. 48640. (silicone products)
9. Eastman Kodak, Distillation Products Ind., Rochester, N.Y., 14603. (layers on plastic film)
10. Harshaw Chemical Co., Div. of Kewanee Oil Co., 1945 East 97th St., Cleveland, Ohio 44106. (the Wick Stick for qualitative TLC)
11. Helena Laboratories, Taylor, Mich. 48180. (standard bore capillaries for TLC spotting)
12. Kontes Glass Co., Spruce St., Vineland, N.J. 08360. (etched glass plates)
13. Krylon, Dept., The Borden Chemical Co., P. O. Box 390, Norristown, Pa. 19404. (plastic spray for embedding thin layers for documentation)
14. Nuclear Chicago Corp., 333 E. Howard Ave., Des Plaines, Ill. 60018. (counting equipment)
15. Pharmacia Fine Chemicals, Inc., 800 Centennial Ave., Piscataway, N.J. 08854. (Sephadex gels)
16. Photovolt Corp., 1115 Broadway, N.Y. 10010. (densitometers)
17. H. Reeve Angel & Co, Inc., 9 Bridewell Pl., Clifton, N.J. 07014. (adsorbents)
18. Rodder Instruments, 775 Sunshine Dr., Los Altos, Cal. 94022. (streaking device for preparative TLC)
19. Schoeffel Instrument Co., 15 Douglas St., Westwood, N.J. 07576. (counting equipment)
20. G. K. Turner Associates, 2524 Pulgas Ave., Palo Alto, Cal. 94303. (fluorescence equipment)

Column Chromatography—Adsorbents

1. Aluminum Co. of America, 1501 Alcoa Bldg., Pittsburgh, Pa. 15219.
2. Alupharm Chemicals, 610–612 Commercial Pl., P. O. Box 30628, New Orleans, La. 70130.
3. Bio-Rad Laboratories, 32nd and Griffin Aves., Richmond, Cal. 94804.
4. Brinkmann Instruments Inc., Cantiague Rd., Westbury, N.Y. 11590.
5. Fisher Scientific Co., 711 Forbes Ave., Pittsburgh, Pa. 15219.
6. Floridin Co., 375 Park Ave., New York, N.Y. 10022.
7. Fluka A. G., Buchs, S. G., Switzerland.
8. W. R. Grace & Co., Davison Chemical Division, 101 N. Charles St., Baltimore, Md. 21203.
9. Johns-Manville Products Corp., Celite Division, 22 E. 40th St., New York, N.Y. 10016.
10. Mallinckrodt Chemical Works, Second and Mallinckrodt Sts., St. Louis, Mo. 63160.
11. H. Reeve Angel & Co., Inc., 9 Bridewell Pl., Clifton, N.J. 07014.
12. Carl Schleicher & Schuell Co., 543 Washington St., Keene, N.H. 03431.

Column Chromatography—Columns

1. Ace Glass Inc., P. O. Box 688, Vineland, N.J. 08360.
2. Fisher & Porter Co., 523 Warminster Rd., Warminster, Pa. 18974.
3. Kontes Glass Co., Spruce St., Vineland, N.J., 08360.
4. Scientific Glass Apparatus Co., Inc., 735 Broad St., Bloomfield, N.J. 07003.

Column Chromatography—Fraction Collectors

1. Buchler Instruments, Inc., 1327 16th St., Fort Lee, N.J. 07024.
2. Gilson Medical Electronics, 3000 West Beltline Hwy., Middleton, Wisc. 53562.
3. LKB Instruments, Inc., 12221 Parklawn Dr., Rockville, Md. 20852.
4. Rinco Instrument Laboratories, Inc., 503 Prairie St., Greenville, Ill., 62246.

Column Chromatography—Gradient Mixers

1. Ace Scientific Supply Co., Inc., 1420 East Linden Ave., Linden, N.J. 07036.
2. Buchler Instruments, Inc., 1327 16th St., Fort Lee, N.J. 07024.
3. National Instrument Laboratories, Inc., 12300 Parklawn Dr., Rockville, Md. 20852.

Gas Chromatography—General Equipment

1. American Instrument Co., Inc., 8030 Georgia Ave., Silver Springs, Md. 20910.
2. Barber-Colman Co., Industrial Instruments Div., 1300 Rock St., Rockford, Ill. 61101.
3. Beckman Instruments, Inc., 2500 Harbor Blvd., Fullerton, Cal. 92634.
4. Burrell Corp., 2223 Fifth Ave., Pittsburgh, Pa. 15219.

5. Central Scientific Co., Div. of Cenco Instruments Corp., 2600 So. Kostner Ave., Chicago, Ill. 60623.
6. Disc Instruments, Inc., 2701 So. Halladay St., Santa Ana, Cal. 92705.
7. Fisher Scientific Co., 711 Forbes Ave., Pittsburgh, Pa. 15219.
8. Hewlett-Packard Co., F & M Scientific Div., Rt. 41, Avondale, Pa. 19311.
9. Jarrell-Ash Co., 590 Lincoln St., Waltham, Mass 02154.
10. Loenco, Inc., 2092 N. Lincoln Ave., Altadena, Cal. 91001.
11. Packard Instrument Co., Inc., 2200 Warrenville Rd., Downers Grove, Ill. 60515.
12. Perkin-Elmer Corp., 702-G Main Ave., Norwalk, Conn. 06852.
13. Varian Aerograph, 2700 Mitchell Dr., Walnut Creek, Cal. 94596.
14. Victoreen Instrument Co., 10101 Woodland Ave., Cleveland, Ohio 44104.
15. Warner-Chilcott, Instruments Div., 200 S. Garrard Blvd., Richmond, Cal. 94804.

Gas Chromatography—Specific Items

1. Analabs, Inc., P. O. Drawer 5397, Hamden, Conn. 06518. (coated supports and packed columns)
2. Hamilton Co., 12440 E. Lambert Rd., Whittier, Cal. 90608. (syringes)
3. Becton, Dickinson and Co., Rutherford, N.J. 07070. (syringes)
4. Crawford Fitting Co., 29500 Solon Rd., Solon, Ohio 44139. (Swagelok tube fittings)

Index

153

Partition coefficient, 30
  definition, 13
Pauson, P. L., 148
Peak enhancement, 131, 132
Pecsok, R. L., 147
Peifer, J. J., 41
Peifer slurries, 43
Penicillins, 38
Perchloric acid, 67
Pereira, R. L., 133
Perisho, C. R., 20
Perkin-Elmer Corp., 152
pH of layers, 55
Pharmacia Fine Chemicals Inc., 46, 150
Phenolic acids, 34
Phenols
  separation, 34, 38
  visualization, 70
Phenylhydrazones, 31
Phillips, C., 147
Phosphors, TLC, 42
Phosphorus detector, 127
Photovolt Corp., 80, 150
Phthalates, 35
Planimeter, 78, 134, 136
Plaster of Paris binder, 42
Plasticizers, 35
Polar compounds, 85, 90, 100, 103
Polarity, 22, 25, 111
  dielectric constant, 22
Pollard, F. H., 147
Polymers, 141
Polyol, 113
Polyvinyl alcohol binder, 42
Poropaks, 120
Prechromatographic fractionation, 85
Preequilibration, TLC, 59, 64
Preparation of layers, 49
Preparation of slurries, 49
Preparative applications, 11
Preparative GLC, 12, 139
  automatic cycling apparatus, 140
  enlarged diameter column, 140
  parallel columns, 141
Preparative TLC, 68
  adsorbent, 71
  applications, 71
  development, 75
  layer activation, 73
  layer preparation, 72
  layer sizes, 73
  layer thickness, 72
  "P" adsorbents, 71
  precautions, 77
  sample application
    commercial gadgets, 73–74
    homemade gadgets, 74–75
  sample recovery, 76
  sample size, 75
  sample spotting, 73
  slurry preparation, 72
  visualization, 76
Pressure drop, 20, 120
Purdy, S. J., 77

Purnell, H., 147
Pyrolytic GLC, 139, 141

Qualitative applications, 8
Qualitative chromatography, fingerprint
  patterns, 9
Qualitative identification, 8, 130
Quantitative analysis, GLC, 8, 11, 134
Quantitative analysis, TLC, 77
  assay on layer, 77
  densitometry, 80
  elution methods, 81, 82
  radioactive methods, 83
  sample spotting, 78
  spot area, 77
Quantitative chromatography, 11
Quickfit Inc., 55, 150

Randerath, K., 145
Reagents in solvent, 58
Recorders, 8, 109, 115, 128, 132
Reducing (diaphragm) valve, 114
Reeve Angel & Co., 50, 101, 150, 151
  adsorbents, TLC, 50
Reichstein, T., 93
Research Specialties, 46
Resolution, 115
Retention time, 8, 9, 123, 130, 131, 132
  definition, 8
Reversed phase chromatography, 99,
  102, 103, 104
  selection of a solvent system, 36
Review articles
  column chromatography, 146
  GLC, 147
  TLC, 146
Reviews of Analytical Chemistry, 144
R_f, 130
  definition, 5, 16, 17
  in TLC, 60, 62, 64
  publication, 65, 68
  sample size, 65
  TLC vs. column, 86
Rice, C. B. F., 145
Rinco Instrument Laboratories, Inc., 151
Rodder Instruments, 72, 73, 150
Rose, B. A., 147
Rothweiler, W., 48⁻

S-chambers, 59, 62
Sadtler spectra, 134
Safranski, L. W., 147
Sample
  injection, 8, 109, 115
  pretreatment, 133
Sample recovery, preparative TLC, 76
Sample size, 58
  effect on R_f, 65
  GLC, 109, 118, 121, 125, 127, 133,
    140
  preparative TLC, 75
Sample splitting, 118